TABLE OF CONTENTS → P. 15

HALF A CENTURY AGO, HUMANS
FIRST SET FOOT ON THE MOON . . .

FROM THERE, AN ENTIRELY NEW VIEW OF EARTH
OPENED UP: SEEN FROM A DISTANCE AND IN HER
COSMIC SURROUNDINGS, EARTH RESEMBLES A
LONELY "BLUE MARBLE" SILENTLY FLOATING IN
AN ENDLESS, ALL BUT EMPTY OCEAN OF SPACE.

→ A view of "Spaceship Earth," as
seen from the Moon, July 20, 1969
during NASA's Apollo 11 mission.
Photo: NASA

THAT NEW VIEW GAVE RISE TO AN AWARENESS
OF JUST HOW UNIQUE AND FRAGILE OUR COSMIC
HOME MAY BE—AND THAT WE HUMANS HAD BETTER
PROTECT IT AS BEST WE CAN. WHILE THAT IDEA
OF "ENVIRONMENTALISM" CONTINUALLY GAINED
STRENGTH OVER THE NEXT DECADES, AND MANY
HAVE FOUGHT THE EXPLOITATION OF OUR PLANET,

← An astronaut's bootprint in lunar
soil, July 20, 1969. Photo: NASA

. . . A GLOBALIZED FREE-MARKET ECONOMY AND
ITS EXTRACTIVE INDUSTRIES HAVE SINCE TIGHTENED
THEIR GRIP ON NATURE AND GROWN MANKIND'S
ECOLOGICAL FOOTPRINT TO GIGANTIC DIMENSIONS.
UNDER THIS REGIME "BUSINESS AS USUAL" MEANS
ONE THING ONLY:

1 Oil platform in the North Sea
Photo: iStock.com / morkeman
2 Surface mining of lignite
(brown coal) in Garzweiler, Germany
Photo: Raimond Spekking (CC BY-SA 4.0)
3 Coal trains in Wyoming, USA
Photo: Kimon Berlin (CC BY-SA 2.0)
4 A Dutch industrial fishing trawler
off the coast of Mauritania
Photo: Pierre Gleizes / Greenpeace
5 Logging for land clearing in Papua,
Indonesia. Photo: Ardiles Rante / Greenpeace

1. TAKE

2. MAKE

6 Steel production
Photo: Magdalena Iordache / Alamy Stock Photo
7 Oil refinery. Photo: Adobe Stock / kapichka
8 Assembly line at a car manufacturing
plant in China. Photo: iStock.com / xieyuliang
9 Baby chicks on a poultry farm
in Massachusetts, USA
Photo: Edwin Remsberg / Alamy Stock Photo

3. WASTE

DURING A SINGLE LIFETIME THE WORLD'S POPULATION HAS MORE THAN TRIPLED, AND MANY OF US NOW PURSUE LIFESTYLES THAT CONSUME FAR MORE RESOURCES THAN <u>ONE</u> EARTH CAN REPLENISH. INADVERTENTLY, MANKIND HAS THUS EVEN LAUNCHED ITS OWN GEOLOGICAL EPOCH: IN THE "ANTHROPOCENE" HUMANS HAVE SURPASSED NATURE AS THE GREATEST FORCE OF CHANGE ON THE PLANET. THIS IS HOW WE LIVE . . .

NOW

HOW I WILL STOP GLOBAL HEATING BY TONIGHT,
AND LIVE SUSTAINABLY EVER AFTER . . .

A VISUAL GUIDE TO
THE SCIENCE AND EVERYDAYNESS
OF THE CLIMATE CRISIS

Christian Schienerl

Verlag für moderne Kunst

FOLLOW ME THROUGH THIS DAY →

HOUR BY HOUR, WE'LL UNCOVER THE
1. SCIENCE BASICS, 2. CAUSES, AND
3. IMPACTS OF THE CLIMATE CRISIS—
AS WELL AS 4. SOLUTIONS TO IT.

IN THE FOUR MY FOOTPRINT SECTIONS
YOU WILL SEE HOW I AM DOING WITH
STOPPING GLOBAL HEATING . . .

Table of contents

IT'S JANUARY 11, 2021—AS GOOD A DAY AS ANY TO TACKLE THE CLIMATE CRISIS, IF YOU ASK ME.

HELLO, MY NAME IS CHRISTIAN SCHIENERL, I LIVE AND WORK IN THE CITY OF VIENNA IN AUSTRIA, AND I HAVE SET MY MIND ON STOPPING GLOBAL HEATING BY THE END OF THIS DAY.

IF YOU THINK I AM NUTS NOW, YOU MAY BE RIGHT. BUT YOU KNOW HOW THEY SAY THAT "NO ONE IS TOO SMALL TO MAKE A DIFFERENCE"?

WELL, WE WILL PUT THAT CLAIM TO THE TEST TODAY. AND I CAN'T THINK OF A BETTER PLACE TO DO SO THAN A RICH COUNTRY LIKE AUSTRIA . . .

→ **The city of Vienna**
In the foreground: the waste incineration plant in the ninth district.
Photo: Wiener Wildnis

I wonder when always having "to be on time" and constantly delivering "just in time" have become the bane of my existence? Photo: Christian Schienerl

I HEAR THREE CLOCKS TICKING—YET I AM ALREADY LATE!

My wristwatch shows 8 a.m. sharp—and that means I should be on my way to the office already. But I'm still brushing my teeth. Somehow I missed the alarm clock this morning. Looking at my tired face in the bathroom mirror, I wonder when always having "to be on time" and constantly delivering "just in time" have become the bane of my existence. But I guess I am not alone with that: Ask anyone these days what they consider the most precious good in their lives—chances are the answer will be "time."

Time is of the essence in this book as well. The many watchfaces throughout its pages are no coincidence. Even its title—*NOW*—already refers to a very special point in time in all of our lives: "Now" is the moment constantly front and center in our minds, and also the only moment we can ever do anything.

While that magic "now-moment" never changes, and will forever remain the only one we can be in, in this book the speed and direction of time's flow may occasionally change. Now my wristwatch shows 8:02 a.m., and I am even less on time for work. That is a case of me progressing *forward* in time, and much too rapidly at that. But there is nothing unusual about that . . .

Besides my wristwatch though, lately I keep hearing another, very big clock ticking. Its second hand thuds and rumbles in the back of my head like distant thunder. This clock is actually ticking for the whole world to hear, and indicates the progress of something happening on a planetary scale: *climate change*, or—more adequately these days—the *climate crisis*. This second clock shows two minutes past twelve, and it will only become louder the longer we listen to it. The climate crisis is also something progressing *forward* in time much too rapidly. First, things have escalated from climate change to climate crisis, and if we just keep listening they will go on escalating, on to an emergency, then on to climate breakdown. Fortunately, we can act NOW to avoid that, slow down the clock, maybe even stop it. We will never be able to reverse that clock though. That is a property only the third clock I hear possesses . . .

Occasionally, when time comes to a halt—during moments when future and past are blissfully suspended—my inner ear can hear an even bigger timepiece pounding: On this third clock, the second hand moves backwards, each tick marking the passing of a million years. T I C K - T O C K—and the last two million years have flashed by. Back then, there are no shopping malls or modern humans on Earth. Our remotest ancestor *Homo habilis* ("handy man") has probably just figured out how to use stone tools.

Go back 4,567 ticks (4.567 billion years) on that third clock and you arrive at the birth of Earth. Now there are no animals, plants, or even microbes present on the planet—no life at all. All there is is an enormous blob of molten rock, thousands of degrees Celsius hot, and incandescent red against the black of outer space. No moon orbits this recently formed planet. But another of the dozens of planets circling the newborn Sun is headed for Earth on a collision course . . . Our planet will survive the crash, and out of it will spring its trusty companion from then on, the Moon. Fast forward almost 4.6 billion years: On July 20, 1969, the first man sets foot on the Moon and looks back at Earth, that lonely "blue marble" floating silently in an endless and cold ocean of space. She has withstood many a grand catastrophe during her multibillion-year history. We can imagine how that man's eyes—probably Neil Armstrong's—fill up with tears as he looks back at his home from a distance of more than 300,000 kilometers.

Now go back even further on the third clock—13,800 ticks in total—and you arrive at the beginning of everything: space, time, matter, the laws of nature, the boundaries of knowledge. They are all born within a tiny instant, oddly named the *Big Bang*. An almost infinitely small, dense, and hot point explodes, and spawns what we know today as the Universe. As distant in time and as unreal as that incident may seem, its consequences can still be measured and felt today. It is in that sense that my third clock can and will run backwards in time—but no further than that odd moment when everything began. This is how far our minds and astrophysics can travel. Anything before and beyond remains a mystery.

Fast forward again to today, January 11, 2021, shortly after 8 a.m. Central European Time. Zoom in on Vienna, Austria, the house I live in, the bathroom in my apartment: I have finally finished brushing my teeth, and we are ready to go!

In the following account of *How I Will Stop Global Heating By Tonight, and Live Sustainably Ever After*, we will hear the three aforementioned clocks ticking simultaneously. We will find out how each of them relates to the other two: how the clock I am on affects the progression and timescale of the climate crisis; how present-day climate disruption affects living planet Earth, which has evolved over 4,567 million years and only moments ago (ca. 250,000 years) welcomed a new species. This species has named itself *Homo sapiens*, "wise man" . . . We will see how appropriate that characterization is further on. But right now the loudest of the three clocks is my wristwatch, and it shows 8:15 a.m. And that means this Sapiens is really damn late for office—so let's get going!

→
DIFFERENT DEVICES
FOR MEASURING AND
DISPLAYING THE
PASSING OF TIME

↑ Analogue wristwatch
Photo: Christian Schienerl
↗ Sundial. Photo: iStock.com / Kalulu
→ Treerings. Photo: iStock.com / pinboke-oyaji
↓ Eleven annual layers of the GISP2 **ice drill core**. These layers are 16,000 years old and were drilled in Greenland from a depth of 1,855 meters. Where on Earth it was cold enough in the past, and fallen snow did not melt during summers, it got condensed into ice layers when new snow fell on it during subsequent winters. Thus, ever thicker ice sheets grew over the land masses of Antarctica and Greenland. The image shows summer ice layers (arrowed) sandwiched between darker winter layers. Historic carbon dioxide as well as oxygen levels in the atmosphere can be measured directly from ice drill core samples. They are icy time capsules from eras long gone. Photo: NOAA

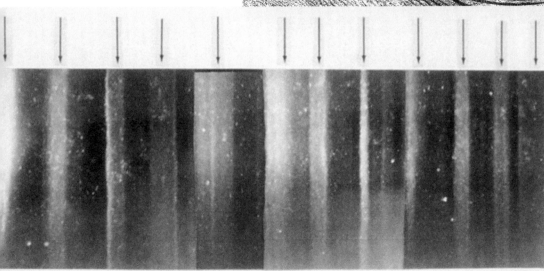

1. Science basics

Earth and the Sun as seen from the Space Shuttle *Columbia*, Jan. 22, 2003. Our solar system with its eight planets orbiting a single central star came into being 4.567 billion years ago. It took just as long for life on Earth to evolve to the complex and burgeoning state we know today. So far, there is not even a hint of another "living planet" like Earth in all of the known Universe. Photo: NASA

It's January 11, the middle of winter, where I live. Accordingly, I put on my rainboots to withstand the dirty slush in the streets. Photo: iStock.com / nycshooter

Why climate and weather matter every day

I step out from the building, and it's a cold morning—as expected. After all, it's January 11, the middle of winter, where I live: Vienna, the capital of Austria, a tiny country in the heart of Europe, on the Northern Hemisphere of the planet. In January, the temperature here is +0.3 °C (Celsius) on average. We live in a so-called *temperate climate*, with four pronounced seasons. Now it is winter, and it should be cold. That is what I learned in school, and that is what I experienced throughout my entire life. So that is what I expected subconsciously when I got dressed a few minutes ago: I put on my thick winter coat, my woolen cap, and boots to withstand the dirty slush in the streets. It may get a few degrees warmer than usual today—I didn't check the weather report—or there may be a snowstorm coming. It may even be sunny all day. But in general, I should be fine with what I wear for the climatic conditions to be expected here on a January day.

As I begin the twenty-minute walk to my office, I think about the difference between *weather* and *climate*. There's a quote by Robert A. Heinlein, which kind of says it all: "Climate is what you expect, weather is what you get."[1]

Weather will always surprise you: sudden drops or rises in temperature (within hours or even minutes), unannounced rain- or snowfall, abrupt swings between cloudiness and sunshine, gusts of wind coming out of nowhere, a freak hailstorm. With today's satellite technology and elaborate software prediction models, meteorological services can give a pretty good look ahead at the weather for a little longer than a week.[2] But any point in time beyond that, and weather just remains a big question mark. The many factors that influence and drive weather simply have not converged yet to form the weather we are going to get. Some of these factors will be mere chance events, but one important and usually reliable driver of weather is the climate it happens in.

Climate is what's to be expected: what the environment is supposed to feel like on any given day of the year in terms of temperature, likelihood of rain- or snowfall, cloudiness/sunshine, windspeed, etc. Climate is the canvas—its shape, size, and material—that weather is going to paint its colorful and expressive image on. Because we humans are so used to the different climates we live in, so intimately connected to them through almost everything we do, it is hard to take even one step back from your climate and see it for what it really is: the way we build, the way we clothe ourselves, the time we have siesta, how and when we plant seeds, how we live through the rainy season, how we survive the dry season—all these immediate human concerns are determined to a large part by the climate they happen in.

As for this expectability of climate, take me as an example: Stepping out from the building on January 11, I do not expect to get all sweaty in my thick winter coat because temperatures have suddenly risen to +20 °C. I also don't expect to freeze to death on my way to the office because a −30 °C cold front has swept over the country over night. Indeed, both these things *could* happen as freak weather events, but in the present climate they are just not very likely. Only a small number of different climate zones exist throughout large parts of the planet (see illustration →). Within these climate zones regional climates or even location-specific *microclimates* exist. Human knowledge of the large climate zones and the climates within them is ancient (meaning a few thousand years old), and it has allowed us to settle into many different lifestyles on planet Earth. We are able to make a living almost anywhere *because* we deal with predictable environmental conditions.

The **predictability of climate** is mostly due to Earth's physical relation to the Sun (see p. 27) and is expressed in statistical terms nowadays. An example: How warm or cold I intuitively expected my environment to feel this morning can be defined exactly as the statistical average temperature of a January day in Vienna. It is calculated from actual **measurements over a period of at least thirty years**. We have had exceptionally warm Januaries during those thirty years, and there will have been surprisingly cold ones. But add those thirty measured temperatures together, divide the result by thirty, and voilà, this will give you the *mean* surface temperature for a January in Vienna. The temperature outliers will have vanished. If you do such calculations for all cities and regions in Austria, you get the **mean surface temperature** for a January in Austria.

You will have heard announcements like "This year's June has been the warmest since the beginning of temperature recordings." Such recordings were undertaken in a systematic fashion from the late eighteenth century onwards. So meteorologists have quite an extensive record to draw on when they compare recent measurements with the known long-term average to arrive at conclusions like the one above. Nowadays, comparisons of present measurements with historic averages draw on more recent base periods though (say 1850–1900), because measurements and recordings were already widespread and highly precise then. Of course you can do such statistics for all months of the year, and for all countries plus the oceans and ice-covered parts of Earth—as the World Meteorological Organization (WMO) or the U.S. National Oceanic and Atmospheric Administration (NOAA) do. That will give you the **global mean surface temperature (GMST)**. The result of present measurements gathered over periods of at least 30 years should conform to the known long-term average, right? Because the physical conditions determining climates have not changed in a very long time. Since the end of the last ice age actually, between 12,000 to 10,000 years before the present, when the interglacial period began, which we are currently in (or were until recently), called the *Holocene* (see pp. 47–48).

But as of late you do not get the expected results anymore when measuring the global mean surface temperature: from a value of approximately +14 °C, GMST has risen by ca. 1 °C during the last 150 years to now +15 °C (see graph p. 31).[3] If you only take Earth's land areas into account, the temperature increase

DIFFERENT climates **EXIST THROUGHOUT LARGE PARTS OF OUR PLANET.**

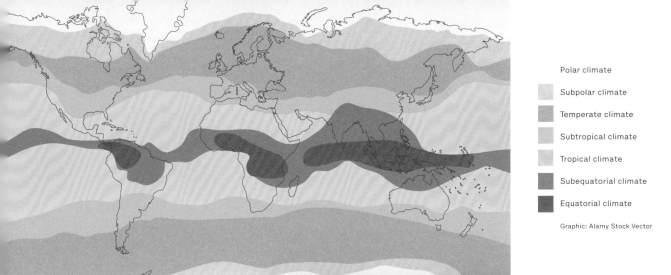

Polar climate

Subpolar climate

Temperate climate

Subtropical climate

Tropical climate

Subequatorial climate

Equatorial climate

Graphic: Alamy Stock Vector

While no such thing as a "global climate" exists, a few distinct **climate zones** (see illustration ↑) span the whole globe like a net. As with an actual net, pulling at some knot is going to affect other knots as well. You can conceive of this net as the **global climate system**.

This system is influenced by many factors, but the most important have to do with **Earth's physical relation to the Sun** (see illustration ↓): Our planet's distance from the Sun on its annual orbit (on average 150 million kilometers) determines the amount of energy Earth receives from our solar system's star. While that amount is basically the same everywhere on the globe (342 Watts per square meter), the angles at which sunrays hit different spots

on Earth (see yellow lines ←) determine those spots' climates. In the *tropical* and *(sub-)equatorial* climate zones (see yellow, orange, and red areas on the map ↖) sunshine is about the same all year round, because sunrays hit at an almost perpendicular angle. The polar regions, on the other hand, receive sunrays at a very flat angle, plus they are facing away from the Sun for half of the year due to Earth's tilt on its own axis (23 degrees, see white dotted line ←). That tilt causes Earth's Northern Hemisphere to have winter season while the southern one has summer.

Photos left and middle: NASA

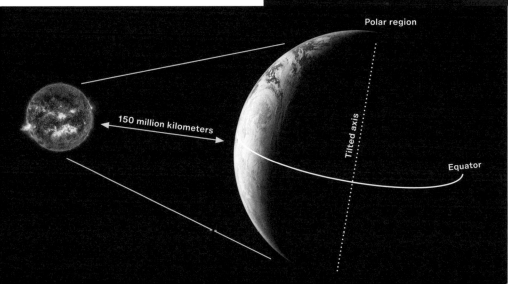

Polar region

150 million kilometers

Tilted axis

Equator

→ The farmer tending this bone-dry field in Germany in August 2018 did not expect the *temperate climate* he lives in to have changed by so much already that an exceptionally long **heatwave**, combined with hardly any rainfall for months, would cost him his crops that year. Photo: iStock.com / ollo

↓ While these children in Ethiopia are used to a hot and arid *subequatorial climate*, the exceptionally long and intense **drought** of 2015 (the worst in decades) forced them to look for and carry home water across many miles of desert plane. As a rule of thumb, the climate crisis makes dry regions of the globe drier and wet regions wetter. Photo: Christine Osborne / Alamy Stock Photo

← This young man on the island nation of Kiribati in the Pacific has to swim through parts of his village that become inundated during the high tide as of late (Photo: Panos Pictures / Vlad Sokhin). This is an impact of **rising sea-levels** due to global heating and thermal expansion of the ocean (water gains volume as it warms) as well as due to

↑ heating-induced **melting of ice masses** in all of Earth's *cryosphere (cryos* = ice). Alpine glaciers and the Greenland and West Antarctic ice sheets are losing mass at unprecedented rates. The image shows an iceberg having calved (broken) off the Greenland ice sheet, floating in Disko Bay, near Ilulissat, July 24, 2015. Photo: NASA

←

BUT ALL OVER
THE GLOBE THESE
CLIMATES ARE NOW
changing . . .

has even been 1.5 °C. In certain regions, like the Arctic, warming has been much higher still, up to 5–6 °C. In Austria, where I live, the increase has been 2.5 °C. This trend of rising temperatures all over the globe during the past 150–200 years is what was called global warming until a few years ago, but now is more aptly known as global heating. Nineteen of the twenty warmest years on record have all occurred during the last twenty years (2001–2020).[4] These are not outliers anymore—this is an unequivocal trend: The planet is heating up.

Where has this additional heat come from? Well, here is one answer: Some 200 years ago, God decided that our planet was a bit too cool at a mean surface temperature of approximately +14 °C—the way it had been for the last 12,000 years.[5] God felt like shaking things up a bit to see what would happen. So He turned up a gigantic heater and waited . . . and voilà, today the whole planet is 1 °C warmer! But how did God do it, you may ask? Did He simply increase *solar irradiance*—the amount of radiant energy reaching Earth from the Sun? Scientific evidence says no: Fluctuations in solar irradiance as a potential *natural forcing* of climate change have been insignificant as a contributor to the meas-ured long-term heating trend.[6] Did God increase volcanic activity on Earth then and thus put more heat-trapping carbon dioxide into the atmosphere? (We'll hear more about that gas further on.) Again: no, the scientific evidence shows volcanic activity's insignificance as a natural forcing of the heating trend.[7]

The answer is much simpler: God let his minions on planet Earth do it, men and women like you and me. In scientific language: "[T]he primary components of external forcing over the past century are human-caused increases in green-house gases, stratospheric ozone depletion and change in atmospheric aerosol content, all reflecting human influence on climate."[8]

Not only has it been proven over and over now that current global heating is man-made *(anthropogenic)*, it has also been proven that 97–99.94% of climate scientists wordwide subscribe to these findings.[9] In other words: the *scientific consensus* on the established fact that human activity is causing global heating is almost a hundred percent! There is really nothing left to doubt, "discuss," or deny anymore about the climate crisis being man-made. And to all the skeptics (or deniers) out there who may be confused about the basic concept: Science is always *nothing but* a consensus—a consensus about what can be factually established by research, when it is conducted employing scientific methods. That is something quite different from a consensus about opinions.

We'll find out a little later on *how* we humans did heat up the planet—but now I arrive at the office and have to fumble for my keys. As expected, I have not frozen to death on my way here but I do sweat a little now—what with all these thoughts about global heating . . .

. . . DUE TO
global heating
CAUSED BY
HUMAN ACTIVITY.

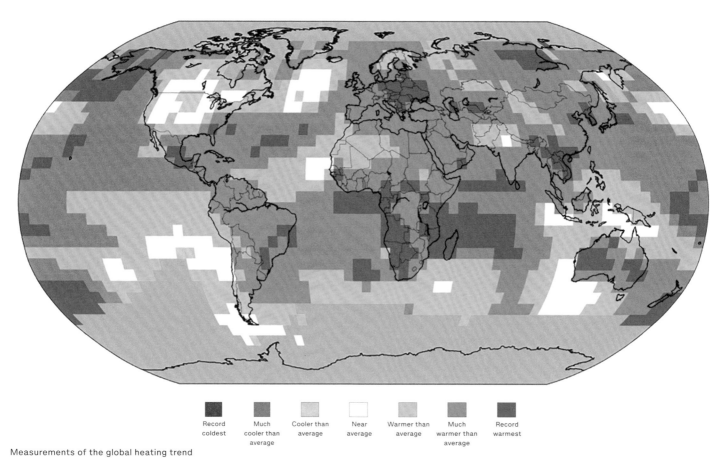

Record coldest · Much cooler than average · Cooler than average · Near average · Warmer than average · Much warmer than average · Record warmest

Measurements of the global heating trend are unequivocal: The map shows **measured land and sea surface temperature distributions for January to December 2019.** Warmer-than-average regions of the globe are colored in light- and mid-red, record-warmest in red. If everything were in order, most of the planet would be near average (white).

Graphic and data source: NOAA National Centers for Environmental Information, State of the Climate: Global Climate Report for Annual 2019, published online January 2020, retrieved May 17, 2020, from: https://www.ncdc.noaa.gov/sotc/global/201913

→ **Increased solar irradiance** (Photo: NASA) and
↓ **increased volcanic activity** (Photo: U.S. Geological Survey) were sometimes proposed as natural forcings of the measured global heating trend in the past. Nowadays, only climate change deniers do so anymore when they are hard-pressed to explain away human activity as the main driver of global heating. Mountains of scientific evidence have long disproved such claims: Both solar and volcanic activity *do* have measurable effects on global temperatures, but these effects have been entirely marginal compared to human-induced heating of the planet during the last 150–250 years (also see p. 29).

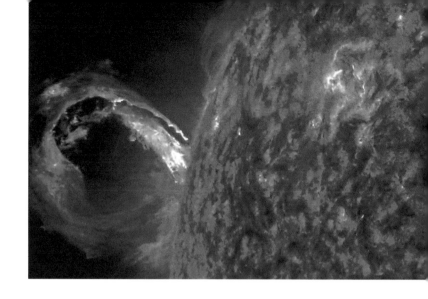

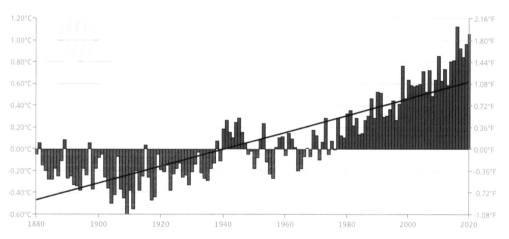

+1.1°C SINCE 1880

↑ **Global land and ocean temperature trend 1880–2020**
The black line in the graphic shows the trend of increasing temperatures between 1880 and 2020 (in total: ca. 1.1 °C; +0.08 °C per decade on average; currently 0.2 °C per decade). Average annual temperature anomalies (blue and red bars) are shown with respect to the twentieth-century average. This is why the black line begins at ca. –0,47 °C and ends ends at ca. +0,62 °C.

Graphic and data source: NOAA National Centers for Environmental information, Climate at a Glance: Global Time Series, published May 2020, retrieved May 17, 2020, from: https://www.ncdc.noaa.gov/cag/

31

A view of the Moon from slightly above Earth's *troposphere*, her lowest atmospheric layer (0–16 km). The troposphere contains all life and the clouds. Photo: NASA

Shielded from the cold of winter by double glass windows and insulated walls, my basil grows just fine in its warm and cozy office environment. Photo: Christian Schienerl

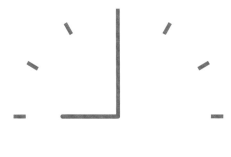

ARRIVING AT THE OFFICE

Atmosphere, greenhouse effect, and "Living Earth"

Walking up the stairs to my office, I realize I forgot to introduce myself properly: So, hi again! My name is Christian Schienerl, I am a freelance book designer, living and working in Vienna, Austria. Now, instead of just doing the layout of books, I have also written one: you are holding it in your hands.

From this introduction you can take away two things: 1. Do not expect this account of a day in my life to be particularly exciting or action-loaded. Although I am a "creative professional" and my job requires a lot of communication, I am surrounded by nothing but books, paper, and printing presses most days. 2. I am no expert in a lot of things I will write about in these pages. But I had some years of academic training (in political science and communications), and I know how to read a scientific paper and what to take away from it. So if my "citizen scientist" discoveries pique your interest, I recommend you delve into my sources referenced in the notes and literature section of this book (from p. 208): These studies, reports, and books represent the current state of scientific knowledge on the various subjects we will touch upon.

So why embark on the adventure of writing and designing a book about the climate crisis and my own role in it, if the issue is not at all my area of expertise? My initial motivation came from growing personal concern and unease about the state of our living environment. What I saw, heard, and read in the media, books, and science papers, but also what I began to perceive and feel changing in my own environment a few years ago made me worry about what may lie ahead for all of us: the future in general, but mostly that of my family and friends, and my personal one. It saddened me to think that we humans as Earth's most recently arrived inhabitants might just carelessly end her intricate beauty and diversity.

Yet there was a huge gap between what I learned was going on out there—how much and how rapidly we are changing our home planet—and my own life realities, which were nowhere near crisis-mode and still produce quite a huge Ecological Footprint each year (see pages 61, 102, 136, 170). This contradiction baffled me. Why is not everyone constantly talking about this seemingly existential threat that we are facing, and acting accordingly? Why don't I?

During the very hot summer of 2018, about the same time I began working on this book, a schoolgirl named Greta Thunberg went on strike in front of the Swedish Parliament every Friday demanding climate protection. In her first public speeches during the fall of that year she asked the exact same question: Why is not everyone deeply concerned about this all the time?[10] Answers to that question are not easily given. I decided to dig deeper and close that knowledge gap myself. I began to make some (easy) lifestyle changes and shared some of the insights I gained with friends, family members, and co-workers. While I registered interest in the subject of the climate crisis—almost everyone agreed that it is *the* most pressing environmental and political issue of our time—almost everyone went immediately back to their regular lives after such conversations and continued doing what they usually do—as if nothing were changing at all. So did I, most of the time, because what can a single person do about it, right? But this "cognitive dissonance" (believing one thing and doing another) kept nagging at me.

After a while, working on the first draft of this book, I realized two things: 1. To almost everyone in my immediate environment, the climate crisis remains something abstract, seemingly far removed from their daily lives and personal sphere of influence. It is a spectre always haunting someone else, most of the time far away (I do not know any farmers or vintners—else the picture might have been different from the start). While this is certainly no longer true—the impacts of the climate crisis are now felt everywhere around the globe, Austria being no exception—it is still relatively easy to dismiss one hottest summer after the other simply as an inconvenience. It is easy to regard each deluge, flood, mudflood, or shrinking glacier as a "natural disaster" like those that have always occurred. It is easy to just quietly and unwittingly adapt to the increasing frequency of such phenomena. 2. Probably closely related to the first reason, a lot of misconceptions regarding the climate crisis prevail, even among highly educated people: the foremost being that a single person cannot do much against the crisis anyhow, and might therefore just as well go on with their lives as usual.

Well, what *can* one do? A mixture of the two aforementioned reasons leads to an affliction I know all too well myself: confusion about what's to be done, a feeling of powerlessness in the face of a seemingly overwhelming problem, and, ultimately, resignation. Whole societies now succumb to such "doom-ism." There is no shame in feeling that way—it's a natural reaction—and somewhat true to the magnitude of the challenge.

But imagine every person and involved player feeling and acting that same way: everyone turning away from the problem and relying on someone else to tackle it. Individuals and scientists relying on politicians to make the necessary regulative changes, politicians relying on industries, technological innovations, and "self-regulating" free markets, nations relying on other nations to lead the fight against climate disruption. After decades of intergovernmental debates on how and when to stop global heating—but nowhere near enough action resulting from such debates—it has become obvious that we are facing exactly that kind of fatal stalemate: Everyone relies on someone else to make the first move, and then no one ever does. In the meantime, environmental conditions have worsened to a point at which NOW everybody would have to make their move immediately, and all in unison. People all over the globe are already losing their livelihoods to changing climates and increasingly frequent extreme weather events. Some nations have even already lost the land they had lived on for hundreds of years to rising seas.

Understandably, Greta Thunberg and upset young people all around the world now demand their future to be protected (**Fridays for Future**, **Extinction**

→
"CLIMATE CHANGE" IS YESTERDAY'S NEWS— TODAY, WE FACE A FULL-BLOWN CLIMATE CRISIS . . .

DER SPIEGEL

3. OKTOBER 1970 · NR. 41
24. JAHRGANG · DM 1,50
INCL. MEHRWERTSTEUER
C 6390 C

Vergiftete Umwelt

DEUTSCHE BUNDESPOST
80
RETTET DEN WALD

November 5, 2012,
on the cover of
*Bloomberg
Businessweek:*

"IT'S GLOBAL
WARMING,
STUPID"

**INDUSTRIES
WARM
THE WORLD**

WASHINGTON, Tuesday.—
Carbon dioxide from man's increased industrial activity was
having a "greenhouse" effect
in the upper atmosphere, tending to make the world's climate warmer, a physicist reported yesterday.

Dr. Gilbert Plass told the
annual meeting of the American Geophysical Union that
the large increase in industrial
activity during the present
century was discharging so
much carbon dioxide into the
atmosphere that the average
temperature was rising at the
rate of 1½deg. each century.

He said that about 2,000,000,-
000 tons of coal and oil were
burnt each year, adding 6,000,-
000,000 tons of carbon dioxide
to the atmosphere.

"The carbon dioxide in the air
acts in the same manner as the
glass in a greenhouse," Dr.
Plass said.

"The carbon dioxide blanket
keeps the ground at an appre-
ciably higher temperature than
it would have if no carbon
dioxide were present."

June 1, 1992,
on the cover of
TIME Magazine:

"**RIO**
COMING TOGETHER
TO SAVE THE EARTH"

Where *is* David Cameron? » | NYC's superskinny skyline »

A week in the life of the world | *Global edition*
15 FEBRUARY 2019 | VOL. 200 No.10 | £4.50 | €6.95*

The GuardianWeekly

I WANT YOU TO PANIC

Climate change meets its match: angry schoolkids.

First warnings about a measurable human influence on the natural greenhouse effect date from the 1950s (see newspaper clipping ↑). My own first experiences with environmental crises date back to the mid-1980s, when the dying forest syndrome from acid rain (see stamp ↑), the Chernobyl reactor disaster, and the hole in the ozone layer were the topics of the day. My initiation to the issue of climate change came with the UN Earth Summit in Rio de Janeiro in 1992 (see cover title of *TIME* ←), which I chose to report on for my highschool graduation exam. It's hard to believe that almost thirty years have passed since then, especially because that "most pressing issue of our time" has become even more pressing in the meantime and turned into an existential threat to mankind.

Global Climate Strike
Edinburgh, UK, Sept. 20, 2019
Photo: Steven Scott Taylor / Alamy Stock Photo

THAT'S WHY MORE AND MORE PEOPLE AROUND THE GLOBE DEMAND THAT THINGS CHANGE NOW!

Rebellion, Sunrise Movement). They act, take to the streets in protest, stop attending school on Fridays to go on strike. They have overcome the insecurity and resignation that have paralyzed all other actors for decades. They demand that things change NOW!

That is why I as well have committed myself to stopping global heating by tonight, and why today I will probe my everyday life for opportunities on *how* to achieve that goal. As the subtitle of this book promises, it will be *A Visual Guide to the Science and Everydayness of the Climate Crisis*. "Everydayness" meaning that this book will "bring the climate crisis home," to my and your doorstep, even beyond our doorsteps, into our homes, and into our lives. There we will then live with it, and it will act like a greedy over-sized pet. We will feed it dozens of kilos of the thing it craves most each and every day. And still, it will want more . . . But relief will come in the form of science and hands-on everyday applications derived from it. I have no doubt: if we can navigate through the basic principles of how this thing called *climate change* works, we will also see how its machinations trickle downwards into our everyday lives. More importantly, we will realize how our daily routines can also trickle upwards to change that thing into something better: better behaved, more manageable—who knows: When we join forces, we might even turn it into something better altogether, something better than what we have now . . .

But first things first: I finally arrive at the office, glad to find it well heated on this cold January morning. We heat with natural gas here, and the thermostat regulating the room temperature is so programmed that it raises the temperature to a cozy 20 °C shortly before I arrive in the mornings. During nights, when no one is here, the thermostat lets the temperature drop to 17 °C. That saves on energy costs, but more importantly, on energy itself.

I make coffee for myself and my assistant Nina, who will also arrive shortly. With a freshly brewed cup of that "black magic infusion" I settle down at my computer to check my e-mails. No need to bother you with that . . . in fact, while reading my mails, even my own thoughts drift off after a while: "We have a nice atmosphere here," is what I think.

Mind you that I only moved to this new office a year ago. Before, I had been working in a space that was practically impossible to heat in winter: hardly any thermal insulation (the building was from the late 1960s), and sitting on the first floor directly above multiple open driveways through the building. Even if you managed to get the air temperature up to normal with the gas heater, the cold of winter would still creep through the floor, into your feet, and further up your body over time.

In comparison, we have a very nice atmosphere here in the new office indeed: heated from above and below (by other apartments in the building) and shielded from the cold of winter by properly insulated walls and double-glass windows . . . And right there it dawns on me: Earth's atmosphere does pretty much the same thing for the planet and for us living on its surface. Although only the thinnest of films covering the massive rocky ball we call our home, its atmosphere completely shields us from the far-below zero temperatures of outer space. Compared to its diameter of 12,742 km, the planet's lower atmosphere with its thickness of a mere 50 km is as thin as the outermost skin of an onion. Yet, this thinnest of layers manages to keep the planet's surface at a cozy +14 °C global mean surface temperature—whereas empty outer space, on which it borders, is –270 °C cold. Exposed to outer space conditions, Earth's surface would immediately freeze. Sure, the surface gets heated from the outside by sunlight during the day, and from below by Earth's hot insides: its molten-rock mantle right beneath the solid rock crust and its molten-metal core even further down (see illustration →). But if it weren't for the atmosphere and something called the natural greenhouse effect (see p. 42) happening within it, all that heat would just radiate out into space, never to be felt again by anyone or warming anything.

You probably know this: walk into a greenhouse, and you'll find yourself in an environment 5–10 °C warmer than outside. How is that? Well, it's an easy-to-grasp principle: the Sun, our solar system's central star, sends out *shortwave lightrays* in all directions. The sunrays hitting Earth during the day warm up her surface and anything on it (including you and me). Anything warmed this way then radiates back *infrared heat energy*. But this infrared heat radiation comes in *longwave* form. Both the glass enclosure of a greenhouse as well as the so-called greenhouse gases (GHGs) in Earth's atmosphere "trap" that longwave heat energy. Neither glass nor GHGs bother with shortwave sunrays—they just let them pass unhindered. But greenhouse gases do like that longwave heat energy emitted back from the surface. So they retain it and do not let it escape to outer space. They also re-emit it in all directions, including back to Earth's surface.

The most abundant of these greenhouse gases is *water vapor*. It accounts for 85% of all GHGs. But as it lasts only a few days in the atmosphere, despite its abundance, water vapor is not the most important GHG. That title goes to the most well-known and talked about of them: carbon dioxide (CO_2). The molecules of this invisible, odorless, and principally harmless gas consist of one atom of the element carbon (C) with one oxygen (O) atom attached to either side of the C atom (hence CO_2). Besides having marvellous heat-retaining and heat-emitting properties, the CO_2 molecule is also pretty sturdy and long-living: once in the atmosphere, it lingers there for up to millenia (!).

Due to these properties of CO_2 and other greenhouse gases like methane (CH_4)—which is twentyeight times more powerful than CO_2 but luckily far less persistent—the natural greenhouse effect has kept Earth's mean surface temperature at a pleasant +14 °C for the last 12,000 years. Without the greenhouse

→
ALTHOUGH ONLY A VERY THIN FILM, Earth's atmosphere PROTECTS ALL LIFE FROM OUTER SPACE'S –270 °CELSIUS.

Relative to Earth's diameter of 12,742 km, her 50 km thick lower atmosphere (seen in the photo as a greenish haze ←) is as thin as the outermost skin of an onion.
Photo: NASA

→ The illustration shows **Earth's atmospheric layers** (fanned and strongly exaggerated): The *troposphere* (**A**, 0–16 km) contains all life on the planet and the clouds. On top of it lies the *stratosphere* (**B**, 16–50 km), whose uppermost layer is made of *ozone* (O_3), followed by the *mesosphere* (**C**, 50–85 km), and the *thermosphere* (**D**, 85–600 km). Outside of the thermosphere lies the *exosphere*—or outer space—with a mean temperature of –270 °C.

The illustration also shows **Earth's inside layers**: the static molten-metal *inner core* (**1**) is enveloped by the *outer core* (**2**), also consisting of metals. In the outer core metals (like iron) are constantly moving due to heat-induced *convection currents*. These metals thereby generate Earth's *magnetic field*, which shields the planet from harmful *solar winds* (streams of charged particles emanating from the Sun). The core is enveloped by the *inner mantle* (**3**), consisting mostly of molten and moving rock. Further out, the *outer mantle* (**4**) produces Earth's very thin *crust* (**5**) where molten rock cools off far enough to solidify: oceanic crust at mid-ocean ridges, and continental crust at so-called *subduction zones*. The very thin layer of the crust is actually moving on top of the outer mantle in several large *tectonic plates*. This regularly leads to reconfigurations of the planet's continental masses during timespans of hundreds of millions of years. This movement of landmasses is sometimes called *continental drift*. The underlying process of *plate tectonics* has played an important role in the evolution of Earth into a living planet. Illustration: Naeblys / Alamy Stock Photo

1

The Sun

emits light in the shortwave blue-green spectrum [A].

2

The atmosphere

(the thin blue haze in the photo) reflects some of the shortwave sunrays [A] back to space.

3

Clouds

reflect some of the short-wave sunrays [A] back to space.

4

Earth's surface

receives a portion of the sun's shortwave rays [A], warms up (the more, the darker the surface is), and emits back longwave heat energy [B] in the infrared spectrum.

5

Greenhouse gases

occur naturally in the atmosphere. 85% are water vapor. The remaining 15% are made up mainly of carbon dioxide (76%), methane (16%), and nitrous oxide (6%). Greenhouse gases let shortwave sunrays [A] pass unhindered, but trap the longwave infra-red heat energy emitted back from Earth's surface [B]. Then they radiate that heat out in all directions [C], including back to the planet's surface—thus creating the greenhouse effect.

6

Humans emit greenhouse gases

into the atmosphere [D], primarily by burning fossil fuels. These man-made emissions add to the naturally occuring greenhouse gases. GHGs linger in the atmosphere for decades (methane) to millenia (carbon dioxide), which is why they have been accumulating there over the past 200 years.

Graphic: SCHIENERL D/AD; photo: NASA

WITHOUT THE NATURAL greenhouse effect **EARTH'S SURFACE WOULD BE FROZEN AT −19 °CELSIUS . . .**

effect, Earth's mean surface temperature would be at −19 °C (!). By the way: The 12,000 lucky years mankind has just lived through were a climatically exceptionally stable *interglacial period*—meaning a period that begins after an ice age and eventually leads up to another one thousands of years later. The present interglacial called the *Holocene* presented humanity with unusually stable Goldilocks conditions, allowing for things like agriculture and civilizations to develop (see pp. 46–48).

The amazing natural greenhouse effect gets even more mind-boggling once you consider the overall **composition of Earth's atmosphere**: It contains only 0.29% greenhouse gases, 85% of which are water vapor, as mentioned before. Therefore, only 0.04% of the atmosphere consist of the other GHGs: carbon dioxide (CO_2, 74% share), methane (CH_4, 16%), nitrous oxide (N_2O, 6%), and 1.8% F-gases (fluorinated gases). Most of the atmosphere is composed of the elements nitrogen (N, 78%) and oxygen (O, 20.9%; a good thing for breathing life-forms). Inert gases like Argon (Ar) make up about 1%.

Should you wonder how an atmosphere works anyhow and how all these gases and gaseous elements in it are kept in place around Earth, there is an easy answer: the same way that *you* are kept in place on the planet's surface instead of flying off into space—yes, gravity! And that Earth gravity (of 1 G) is exactly the right amount to do that, too. Remember this from basic physics? *An object's gravity* (force of attraction) *is proportionate to its mass*. Our neighbour planet Mars would remember, if only we could ask it: Like Earth, Mars also once had an ocean of water on its surface and an atmosphere. But being only a tenth of Earth's mass, Mars's gravity wasn't strong enough to keep both atmosphere and ocean in place. They just evaporated into space over time.[11] Poor Mars! So close to us, yet so astoundingly different from Earth.

But back to the small fraction of greenhouse gases in the atmosphere (0.29%) having such a huge effect: keeping Earth's surface so much warmer than freezing outer space. Can you imagine what temperature difference even a comparatively small rise in atmospheric greenhouse gas concentration would bring about? Unfortunately, this is not something we have to imagine—it has already happened: At the beginning of the industrial era in the 1780s, the **concentration of carbon dioxide in the atmosphere** was at 277 parts per million (ppm).[12] That means that out of a random million air molecules in the atmosphere only 277 molecules were CO_2. Since then, that concentration has gradually risen to currently 415 ppm (2020). Still a very small number compared to a million. But compared to the pre-industrial level of 277 ppm, the current 415 ppm equal a 50% rise! Check the graph on pages 46–47 if you please: You can see that the natural variation of atmospheric CO_2 levels has always been between 180 and 280 ppm for the last 400,000 years (actually the last 800,000 years even). That means something unprecedented has happened over the last 200–250 years: yes, the industry of man and the contemporary civilization we have built for ourselves, fired by seemingly endless and dirt-cheap fossil fuels.

Luckily, the relation between the abruptly risen CO_2 concentration in the atmosphere and increasing temperatures is not a linear one.[13] Otherwise, we would already live—or rather not live anymore—in a catastrophically over-heated world with a mean surface temperature of +21 °C, and summer temperatures of +60 °C in a then not-so-temperate climate anymore. Have you ever been to a desert and tried doing anything in baking-hot midday temperatures?

For now, let's just say we are lucky that the relation between greenhouse gas concentrations and surface temperatures is not linear. Later on, we will find out how intricately complex the relations between different elements in the climate system and the Earth System as a whole actually are, and how negative (dampening) natural feedbacks and positive (reinforcing) feedbacks may completely upend simple linear relations or progressions (pp. 164–169). In the present that means luck for us—in the near future it could turn into the exact opposite . . .

Right there my deliberations come to an abrupt end, as my assistant Nina walks through the door, her nose and cheeks red from the transition between the cold winter morning and the well-heated atmosphere of the office with its 20 °C room temperature.

. . . BUT DUE TO HUMAN ACTIVITY, THE GREENHOUSE EFFECT HAS UNNATURALLY INTENSIFIED OVER THE LAST TWO CENTURIES (SEE NEXT SPREAD).

Biosphere 2 in the Sonora Desert in Arizona was an experimental simulation of the Earth System (see p. 51) on a small scale. It consisted of several greenhouses and buildings connected to each other but completely sealed off from the outside (only sunlight was allowed to enter the complex). The buildings contained an "ocean," a desert, a rainforest, etc. *Biosphere 2* was meant to function as a self-contained ecosystem— just as Earth does (the "Biosphere 1"). But in contrast to its natural inspiration the *Biosphere 2* experiment failed twice within three years and was given up in 1994. Today, the complex of buildings is used for research purposes by the University of Arizona.

Photo: Pat Eyre / Alamy Stock Photo

400,000 years
of natural climate change...

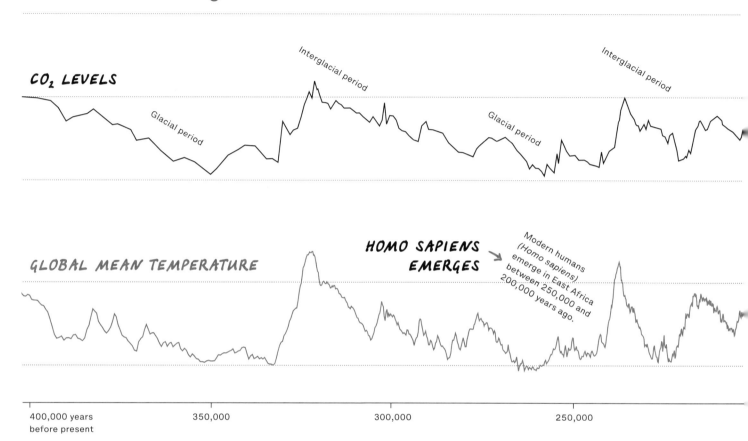

CO₂ LEVELS

Interglacial period

Glacial period

Interglacial period

Glacial period

GLOBAL MEAN TEMPERATURE

HOMO SAPIENS EMERGES

Modern humans (Homo sapiens) emerge in East Africa between 250,000 and 200,000 years ago.

400,000 years before present

350,000

300,000

250,000

Graphic: SCHIENERL D/AD; Data source/source
material: Vostok ice core data (see PETIT *et al.* [1999])
and NOAA Mauna Loa CO₂ record; https://climate.nasa.
gov/evidence/

...versus 220 years of man-made climate disruption

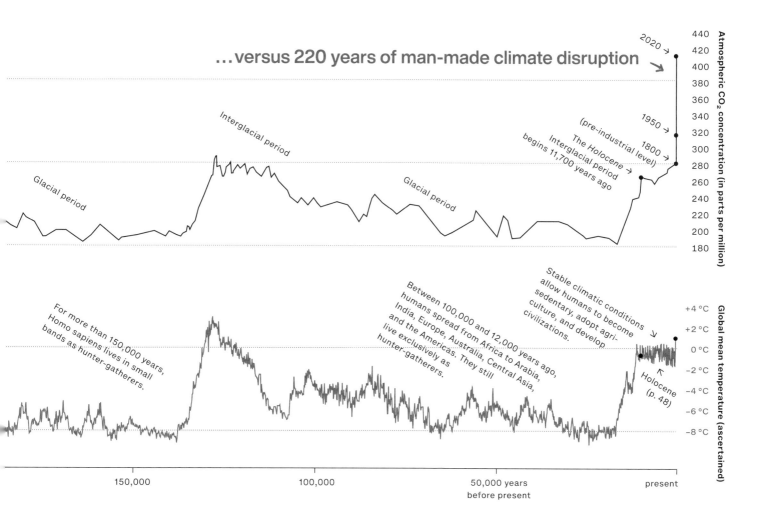

The black graph above shows the varying **concentrations of carbon dioxide (CO₂) in Earth's atmosphere during the last 400,000 years**. The data stems from direct measurements of CO_2 levels in atmospheric samples (tiny air bubbles) contained in an ice core drilled in East Antarctica to a depth of 3,623 meters in 1998 (the Vostok ice core; see PETIT *et al.* [1999]).

As the graph shows, CO_2 concentrations have always varied during those 400,000 years, as did the global mean surface temperature. The orange graph shows mean temperature ascertained to high precision through *isotope fractionation* of ice core water, which is a temperature-dependent process[14].

The natural variability of CO_2 levels and temperatures and the resulting alternation of glacial periods ("ice ages") and interglacial (warm) periods stem from periodical changes in Earth's orbit around the sun *(orbital eccentricity)* in 100,000-year-long intervals, called *Milankovitch cycles*. For the past 1.2 million years, Earth has been going through such cycles, and with them, her climate has always been changing (the graph covers four such glacial/interglacial cycles). Yet the data also shows that natural variability of CO_2 concentrations fluctuated between ca. 180 and 280 parts per million (ppm) in the past (compare black dotted lines).

In stark contrast, CO_2 levels have increased sharply over this natural range since the industrial revolution (see far end of the black graph ↗): While they were at 277 ppm in the late eighteenth century, they have risen to 413–415 ppm in 2020. CO_2 levels are now higher than anytime during the past 800,000 years[15] (that is how far back the con-

tinuous CO_2 record of the EPICA Dome C ice core reaches, which was also drilled in East Antarctica). Not only is a *magnitude* of CO_2 increase as the present one unique—even more so is the *rate* at which it occurs: During the last two decades alone, the rate of increase has been around 100 times that of the maximum rate during the last deglaciation (the rise between ca. 18,000 and 12,000 years before present). No natural driver of CO_2 concentration (like volcanic activity) can cause such a sharp increase. *Carbon isotope signatures* unequivocally mark that sharp rise during the last 200–250 years as stemming mostly from the burning of fossil fuels. In other words: the current rapid increase of CO_2 in the atmosphere has been proven beyond a doubt to be caused by human activity.

For the last 10,000 years . . .

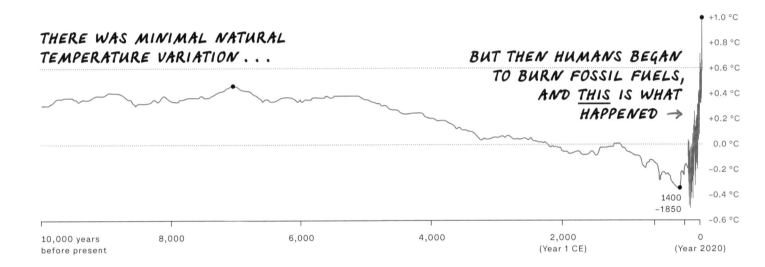

THERE WAS MINIMAL NATURAL
TEMPERATURE VARIATION . . .

BUT THEN HUMANS BEGAN
TO BURN FOSSIL FUELS,
AND THIS IS WHAT
HAPPENED →

+1.0 °C
+0.8 °C
+0.6 °C
+0.4 °C
+0.2 °C
0.0 °C
-0.2 °C
-0.4 °C
-0.6 °C

1400
-1850

10,000 years
before present
8,000
6,000
4,000
2,000
(Year 1 CE)
0
(Year 2020)

For the last 10,000 years, **global mean surface temperature** stayed well within a range of 1 °C (compared to temperature averages during the nineteenth century; see +0.45 °C highpoint at around 7,000 years before present and –0.35 °C lowpoint at around 1850 CE). Historic temperatures (orange graph) are inferred to high precision from ice core samples (see p. 47), and backed up by treering, lake, and sea sediment drill core data. While GMST increased from the end of the last ice age (ca. 18,000 years ago) onward into the Holocene interglacial period until ca. 7,000 years before present, it has been slightly decreasing again since then because we are now in the *cooling phase* of the Holocene interglacial.

The pronounced temperature decrease between 1400 and 1850 (of ca. 0.2 °C) has become known as the *Little Ice Age*. That cooling trend would have gone on for the next few thousand years until Earth would have entered the next glacial period (aka ice age). But that has now changed entirely because

of what happens at the end of the orange curve: The whole graph, with its sharp rise at the far end, is known as the *Hockeystick curve* because this is what its form resembles. It was first presented by climatologist Michael E. Mann and others in 1999, and was widely contested at first. Since then it has been proven to be correct by many new measurements. The Hockeystick curve clearly shows manmade global heating due to additional greenhouse gas emissions since the beginning of the industrial era. Whereas Earth had been on a natural cooling trend since 7,000 years before present, within just 200 years human-caused global heating has pushed temperatures to their highest levels in 120,000 years[16] (for that, see the last pronounced peak on the orange graph on p. 47). Thereby mankind has also very likely averted the coming of the next ice age—which is not good news at all.

Data points from 1880 until present are actual temperature measurements by NOAA. For a less compressed view of those and data sources see p. 31.

Photo: NASA

The Earth System, or "Living Earth"?

Although contemporary human civilization with its cultures, economies, and technologies has tried hard to emancipate itself from the natural world, today we find that the two are not only heavily intertwined but have actually become (or always were) *one* world. Scientists call this one world the **Earth System (ES)**: "The Earth System consists of a suite of interacting physical, chemical and biological global-scale cycles and energy fluxes that provide the life-support system at the surface of the planet."[17] Mankind has not only become part of these interacting cycles and energy fluxes (so-called *biogeochemical flows)*—unwittingly or not, humans are now changing these planetary-scale processes (see pp. 150–155) and by doing so endanger Earth's life-support system. This system has evolved over billions of years into a state of such intricate complexity and interdependency between its elements that it has been likened to a living organism, "Living Earth" or *Gaia* (named after the primordial Greek goddess who personifies Earth). The *Gaia hypothesis* was widely contested when chemist James Lovelock and microbiologist Lynn Margulis put it forth in the 1970s[18] but has been accepted as a scientific theory in the intervening decades. It proposes that living organisms interact with their inorganic surroundings on Earth to form a synergistic and self-regulating, complex system that helps to maintain and perpetuate the conditions for life on the planet. For a comprehensive contemporary survey of the emergent *Systems View of Life* see Fritjof Capra and Pier Luigi Luisi's book of the same name (Capra/Luisi [2016]). Today's *Earth System Science* reflects the Gaia hypothesis in ever-growing knowledge of Earth's different *spheres* and how they interact with and depend on each other. These spheres are:

- The **atmosphere** (see pp. 40–45).
- The **biosphere** is made up of all ecosystems on Earth—terrestrial (land-based), freshwater, and marine (sea-based)—and includes all life-forms *(biota)* living within these ecosystems: animals, plants, funghi, protists, bacteria, and archaea.
- The **hydrosphere** contains all of Earth's oceans, rivers, and lakes, as well as the *cryosphere*. The **cryosphere** is made up of all the ice accumulated at the polar caps, the ice sheets in Greenland and Antarctica, and alpine glaciers all around the world.
- The **lithosphere** is made up of all of Earth's rocks at her surface, in her crust, and in her molten-rock mantle (see p. 41).

Human influence on Living Earth and her spheres has become so pronounced over the last two centuries that the geological epoch of the *Holocene*—the climatically stable interglacial period having begun 11,700 years ago—has highly likely already ended, and we now live in the **Anthropocene**—a new geological epoch in which mankind has surpassed nature as the single most important driving force within the Earth System. The concept of the Antropocene was introduced in detail in 2007 by Will Steffen, Paul J. Crutzen, and John R. McNeill (see STEFFEN/CRUTZEN/McNEILL [2007]). The planetary-scale human influence they described can be seen at work on the *natural carbon cycle*, for example (see pp. 52–55).

↑ With this famous illustration from 1807 (known as "Tableau Physique"), **Alexander von Humboldt** (1769–1859)—outstanding naturalist and explorer, science pioneer, and a trailblazer for modern ecology, climate science, and Earth System science—tried to represent the altitude-dependent vegetation on Mount Chimborazo in the Andes. At the same time, the image represents all of Earth's spheres: atmosphere (**1**), cryosphere (**2**), biosphere (**3**), hydrosphere (**4**), and lithosphere (**5**). Probably no coincidence, as Humboldt was convinced that "everything is connected to everything else."

Image: Science History Images / Alamy Stock Photo

The natural carbon cycle

I am into my third cup of coffee this morning when I notice something strange: Although I have just been sitting for 45 minutes straight, going through the tiniest of corrections on a book we have to deliver to the printers this morning, I feel flushed. I touch my forehead—maybe that sweating in the cold on the way to the office . . . ? But no, it's not a fever. "Is it hot in here, or is it just me?" I ask Nina, who sits at the desk across from me. "Yeah, it's a little warm in here, I was also wondering," she replies. I get up to check the thermostat, which usually displays the current room temperature, but now only shows two ominous dashes: -- °C. I feel the radiator in the room but immediately withdraw my hand—it's blazing hot! "Ouch!" Nina startles at my shriek. "Damn, there is something wrong with the gas boiler. You could make a grilled-cheese sandwich on that radiator," I grumble and go to check the gas boiler in the kitchen. But its display also just shows the two ominous dashes. I unplug the boiler from the wall socket, plug it in again, and wait for it to—what, reboot? Seriously? This thing has a computer inside? When it's done with its routine, the display still just shows the two dashes. "Uh-oh, this thing needs a doctor." "Really? Already? Didn't you have it installed only last year?" Nina delivers my exact thoughts.

So I call the emergency number of the plumber company, and open a window while being on hold. When I finally get to talk to an installer, he advises me to unplug the thing until they arrive. That might be a while though, they are having a lot of emergencies today. Of course.

I open more windows until the room temperature feels kind of normal again with cold air entering from the outside. Back at my desk, I finish the corrections on the book and send the corrected pages off to the printers. No need to worry about the gas boiler for now—the doctor is on the way. With some time to spare, I come to think of something I recently read: how the element carbon (C) is not only the stuff most organic life is made of, but also what that life likes to feed on and draw its energy from (in the form of sugars and carbohydrates). Carbon is also what humans make building materials, glues, paper, all kinds of synthetic fibers (e.g. polyester) and a lot of other stuff from.[19] And of course it is the very strong bonds between individual carbon atoms that make fossil fuels such a great source of energy when burned (pp. 72–89).

Then there is also something called the natural carbon cycle, a mighty set of flows of the element carbon through reservoirs in all of Earth's *spheres* (see p. 51): from the *atmosphere* into the *biosphere* (see pp. 58–63), into the *hydrosphere*, and into the *lithosphere*.[20] Some of these carbon flows occur on a timescale of days, others take millions of years to complete a cycle. The most powerful of these flows is such a painstakingly slow process (see p. 55). It draws carbon from the atmosphere, where it exists in oxidized, gaseous form, namely as carbon dioxide (CO_2). When CO_2 is washed out of the atmosphere by combining with rainwater (H_2O) into *carbonic acid* (the thing that makes the bubbles in your bubbly water), and rains down on rocks on Earth's surface, it makes those rocks *weather*. That chemical weathering produces calcium bicarbonate, which eventually enters the ocean and gets deposited at the sea bottom as carbonate rock sediments. Over long periods of time, these carbonate rocks get drawn into Earth's crust through the process of *plate tectonics*, and from there deeper into the molten-rock mantle. Eventually, this molten carbon will be ejected again from the Earth's insides through volcanic eruptions (on land and underwater). Outgassing happens when the ejected carbon oxidizes again into CO_2 in the atmosphere's air. From there, the cycle will repeat.

This process does not sound particularly spectacular, especially since it takes millions of years to complete—but its effects sure are: This flow of carbon from the atmosphere into the lithosphere and back into the atmosphere works as **Earth's own heating thermostat** (!) Here is how: When we looked at the natural greenhouse effect before (p. 42), you might have raised a tricky question: If carbon dioxide lingers in the atmosphere for millenia, and ever more of it ends up there due to natural processes such as volcanic activity as well as human-caused increases in CO_2 concentration, then shouldn't Earth just get increasingly hotter all the time? And your conclusion would be absolutely right . . . except it doesn't take into account the cycling of carbon through Earth's spheres. In fact, the warmer it gets on the planet, the more CO_2 rains down, and the quicker rocks will weather, because warmer temperatures lead to more water evaporating from the oceans and consequently to more precipitation. That way, more CO_2 gets drawn from the atmosphere into the lithosphere. With less CO_2 in the atmosphere now, Earth cools. When the carbon ends up in the atmosphere again, being shot up there through volcanic eruptions, CO_2 concentration rises again, and Earth warms again. But only up to the point when more and more rocks weather more quickly because a warmer world sees more rain. So Earth will cool again, and the cycle will repeat. The *thermostat* does its job: It regulates the temperature of an enclosed space (all of Earth) by constantly measuring temperatures and increasing or decreasing the flow of the heating element (greenhouse gases). This cycle takes millions of years to complete but it is quite impressive how nature has become its own clever engineer there, isn't it?

Since the thermostat at my office currently does not do its job, and I had to power down the whole heating system, slowly but surely it starts to get cold. Lucky us we still have the old stove, which the space was heated with in the olden days, and also some wood left over. I haven't once used it since we moved in, but how hard can it be to get a fire started?

THE NATURAL carbon cycle **PROVIDES EARTH WITH HER OWN HEATING THERMOSTAT. IT WORKS ON A TIMESCALE OF MILLIONS OF YEARS THOUGH.**

Warming phase

1 CO_2 is outgassing from terrestrial and marine volcanoes. Photo: U.S. Geological Survey

2 More CO_2 in the atmosphere warms Earth (image: model of a CO_2 molecule). Photo: iStock.com / JC559

3 In a warmer world, there is more water vapor in the air (7% more per degree Celsius). Hence, more CO_2 and H_2O rain down on Earth as *carbonic acid*. Photo: iStock.com / olaser

2

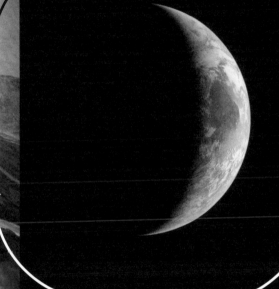

1

6

5

4

Cooling phase

4 As more carbonic acid is drawn from the atmosphere and rains down on Earth's surface, rocks weather more quickly (image). That produces *calcium bicarbonate*. Photo: iStock.com / Elena Estellés

5 Calcium bicarbonate gets deposited in the seas as carbonate sediments, which will form *carbonate rock* (what the fish in the image is hiding under). Photo: NOAA

6 Carbonate rock gets drawn Into the Earth mantle eventually, where it is melted and, in time, will be ejected again through volcanoes as carbon-containing magma. Photo: NASA

Photo center: NASA

55

Amazon rainforest, Ecuador
Photo: iStock.com / Kalistratova

A tree continually grows throughout its lifetime, building new tissue, which can be seen as new treerings forming on its trunk over the years. Just like humans and animals, plants have a *metabolism*, and it turns out we all eat the same stuff to draw energy from and grow: carbon.

Photo: Adobe Stock / guruXOX

Photosynthesis and natural carbon sinks

Indeed, I manage to get a fire going within minutes. The relatively small space of the office warms up rapidly, and while one might get all romantic about the crackling of the fire, I try to understand what is happening inside the stove. Staring into flames will sometimes do that to you: you get all day-dreamy and reflective . . . But I am not pondering the purpose of life in general, or of mine in particular. I am interested in the purpose of the special life-form that has created the very wood, which I am burning now: a tree.

Like other plants, trees seem to have a rather relaxed lifestyle. They don't move around, they always know where their roots are, and they stick with them. While these roots draw nutrients and water from the ground they have grown into, the tree does something different above ground. With its green leaves (or needles)—most of which grow anew each year on its branches—the tree draws two more elements from its environment, which are vital for its growth and survival: the sun's energy, and a gas found abundantly in Earth's atmosphere—carbon dioxide. The tree's metabolism works through the process of photosynthesis: The freely available energy of the sun *(phos* = light) is harnessed by light-absorbing chloroplasts in the tree's green leaves (the chloroplasts produce the natural green dye *chlorophyl*, which gives plants their characteristic color). The chloroplasts convert that solar energy into chemical energy, which is then used to convert another freely available material—carbon dioxide gas—into more energy-rich organic molecules like *carbohydrates*. In the first stage of the photosynthetic process, that chemical energy is used to split the CO_2 molecule into its constituent atoms, namely carbon (C), and a totally superfluous waste-product called free oxygen (O_2). At least to the tree oxygen is entirely superfluous: because it does not breathe. For us breathing beings—humans and other mammals, fish, etc.—oxygen is entirely necessary for survival. Let's just say we are lucky that trees and other plants do not need that oxygen for themselves, and instead just release it into the atmosphere. What trees *do* want though is that delicious and versatile carbon atom. In the second stage of photosynthesis, they therefore synthesize that carbon with hydrogen (H) from the water (H_2O) they have drawn from the ground into *carbohydrates*. Among these carbohydrates is *glucose*, a sugar much like the one we use to sweeten things. And trees do like that glucose! They consume it, and grow new stuff from it: the tissue and fiber of their leaves, branches, and trunks—the wood which I am burning right now, which is just dead plant tissue. All of it is made of carbon, which the plants have drawn from the atmosphere. Now guess what happens when I burn that wood in my stove. The highly energy-dense bonds between the individual carbon atoms, which the tree has stored in its trunk, break up when heated to a certain point, and release that energy, which in turn fuels the fire and keeps it burning. The hot flames give off that heat energy to the surrounding air and that keeps Nina and me warm right now. The now freed and single carbon atoms rise up from the flames and look for new partner atoms to bond with. If they find only a *single* oxygen atom in the air, they bond with it into carbon monoxide (CO), a highly toxic gas for any breathing being. Don't hold your head into the smoke rising from a fire! Most of that smoke is *particulate matter* (PM)—nasty *aerosols* like *black carbon*—which will clog up your lungs, plus carbon monoxide, which will prevent your blood from carrying oxygen through your body.

If the freed carbon atom finds *two* oxygen atoms to bond with, they unite into carbon dioxide again, and that gas will rise to the atmosphere from whence it came when the tree "sucked" it out of the air. That, by the way, is why they say that wood is *carbon-neutral:* when burned, it releases exactly the same amount of carbon back to the atmosphere as the tree originally drew from it through photosynthesis. When the tree drew that CO_2 from the air, it thus became a carbon sink (a *sink* is a reservoir where something is stored and is thus no longer part of a cycle). And that happens every year anew: Old and new plants and trees draw carbon dioxide from the atmosphere at massive levels. Of the roughly 40 billion tonnes of CO_2 we humans currently blow into the atmosphere every year, only 40–45% stay there. The rest is drawn into the biosphere (by plants and other photosynthetic creatures, see Keeling curve →) and into the ocean (which, unfortunately, acidifies increasingly due to higher CO_2 concentrations).

Plants have inherited the photosynthetic metabolism from our farthest-removed common ancestors. As early as three billion years ago, when the only lifeforms on Earth were *bacteria* and *archaea* (another domain of microorganisms), some bacteria discovered the trick of photosynthesis (probably by chance). That trick proved extremely useful though in terms of evolutionary development. There was a lot more CO_2 in the atmosphere back then, and sun energy was also plentiful. Thus, the new model of metabolism became a huge success—so successful in fact, that these photosynthetic *cyanobacteria* have survived essentially unchanged to this day. The first other life-forms to copy that success model were sea-based algaea, called *phytoplankton* (see image →). Much later followed land-based photosynthesizers: mosses and plants.

While the cycling of carbon through biosphere and atmosphere makes wood a renewable energy source, something entirely different and non-renewable emerges when dead plants and animals are being buried deep underground for very long times . . . as we will see in the chapter about fossil fuels (pp. 72–89).

→

Photosynthetic lifeforms

GOBBLE UP A LARGE PORTION (25–30%) OF MAN-MADE GREEN-HOUSE GAS EMISSIONS.

← **A tree** draws nutrients and water from the soil through its roots. Above ground, its leaves, containing tiny **chloroplasts** → capture the energy of sunlight, and convert it into chemical energy. That chemical energy is used to break the CO_2 molecule—also captured by the leaf—into its constituents: carbon, which the tree uses for biofuel and for building its tissue, and oxygen, which the tree releases into the atmosphere as a superfluous waste-product. Photo left: Christian Schienerl; Photo right: Kristian Peters / Fabelfroh (CC BY-SA 3.0)

↓ Photosynthetic **phytoplankton**— so-called *diatoms*—photographed by a scanning electron microscope (Photo: NOAA). In contrast to this microscopic view of phytoplankton, pages 62–63 show a perspective on it from Earth's orbit . . .

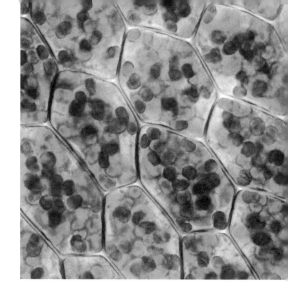

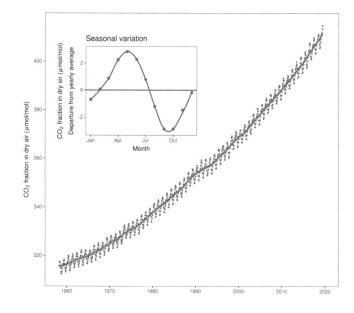

↑ The **Keeling curve** shows the increase of CO_2 levels in the atmosphere since 1958 (given in parts per million [ppm] or "mole fraction" in dry air), when Charles D. Keeling began measuring them at the Mauna Loa Observatory in Hawaii. The curve also shows the seasonal variation in CO_2 concentrations (red zig-zag graph), which is due to increased photosynthetic activity by plants, algaea, and cyanobacteria during the Northern Hemisphere summer season.

Graphic: By Delorme. Data source: Dr. Pieter Tans, NOAA/ESRL and Dr. Ralph Keeling, Scripps Institution of Oceanography

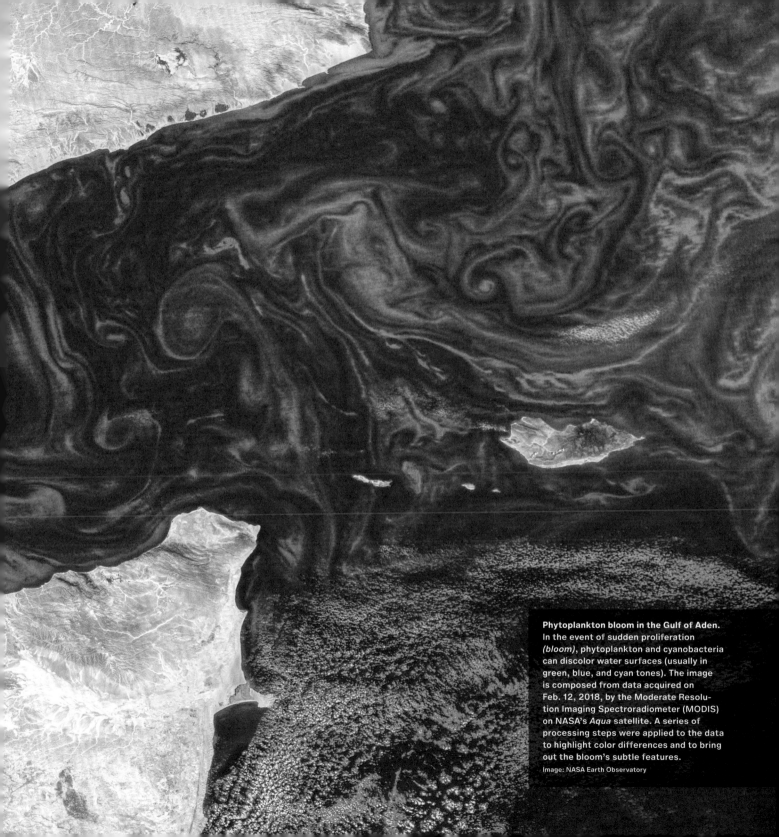

Phytoplankton bloom in the Gulf of Aden. In the event of sudden proliferation *(bloom)*, phytoplankton and cyanobacteria can discolor water surfaces (usually in green, blue, and cyan tones). The image is composed from data acquired on Feb. 12, 2018, by the Moderate Resolution Imaging Spectroradiometer (MODIS) on NASA's *Aqua* satellite. A series of processing steps were applied to the data to highlight color differences and to bring out the bloom's subtle features.
Image: NASA Earth Observatory

The construction of an *Ames Room* creates an optical illusion, which makes similarly sized objects appear to be of different sizes. When we compare the Ecological Footprints of different types of people here (A, B, C), something similar happens: These people do leave footprints of vastly different sizes although they are themselves of similar build and height. Photo: source unknown

A Footprint of Austrian citizen, on average

B Footprint of Earth citizen, on average

C One-Earth-compatible Footprint

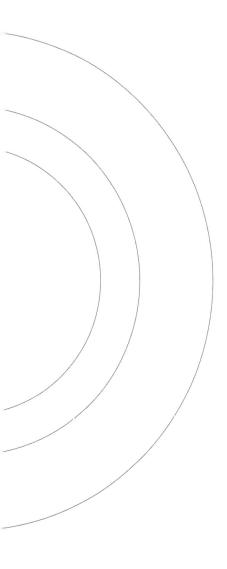

WHAT ARE MY PERSONAL IMPACTS ON THE WORLD? AND WHAT IS AN "ECOLOGICAL FOOTPRINT" ANYHOW?

Regard the three circles to the left. They show differently sized Ecological Footprints of **A** an Austrian citizen, on average, **B** an Earth citizen, on average, **C** what would be a per capita use of resources compatible with *one* planet Earth for all of today's 7.8 billion humans. It might not be obvious, but A is 3.5 times the size of C. That means: **an Austrian citizen requires the resources of 3.5 Earths, on average.** But how is that even possible, as there is only *one* Earth?

Over the course of this day, we will find out where among circles A, B, and C I will have to draw my own, and what it will be filled with.* But one thing is certain from the start: To keep my promise and stop global heating by tonight, I need to touch down within the innermost circle C. I will have to reduce my own contribution to the climate crisis to a level that *one* planet Earth can cope with: by consuming less energy, food, goods, and services, all of which produce the greenhouse gases that drive global heating.

But how will I know which level of resource consumption is *sustainable* in that sense? And what does "sustainability" mean anyhow? Is there an explicit measure for it? Yes, there is, and you have for sure heard of it: the **Ecological Footprint (EF)**. The concept is simple enough to understand: Just as my footsteps leave physical marks on the ground I walk over (my footprints), my sheer being as an organism which consumes all kinds of things leaves an Ecological Footprint on its environment. To survive and go on being I require food, clothing, and shelter, draw on energy for heating, electricity, and mobility, buy and use goods and societal services. All these things have measurable physical impacts on the environment as their production and provision (and disposal) require raw materials, energy, labor, and incidentally lead to byproducts like the emission of greenhouse gases. Though the concept of the Ecological Footprint is easily understood, its actual calculation requires a lot of measuring and accounting of different real-world resource uses. One major part of the Ecological Footprint of a person (or a product, or a city, or a nation), which a lot of contemporary discussions are focussed on, is their **carbon footprint**. Once you have done a lot of measuring and accounting of carbon emissions, you may arrive at a simple statement like "the carbon footprint of an Austrian citizen, on average, is 6.9 tonnes of carbon dioxide emissions per year." That may sound like a lot to you—or not. The figure by itself really does not mean much. It only begins to make sense once you compare it to similar figures like "an American's carbon footprint is 15.5 tons," or "a Congolese's is 30 kg (!)." Or once you relate those individual numbers to

* To calculate your own Ecological Footprint and see how it matches up against mine, visit: www.footprintcalculator.org.

overall emissions: "In 2019, global human-caused CO_2 emissions amounted to ca. 37 billion tonnes,"[21] or "worldwide, we have 320 billion tonnes of CO_2 emissions left to spend for a 66% chance to limit global heating to 1.5 °C."[22] Now, *that* affords some perspective: 320 billion tonnes remaining *carbon budget* divided by almost 40 billion tonnes emissions per year give us roughly eight more years of "business as usual." Then time will be up, we will have spent our entire budget— and we would have to switch to a *decarbonized economy* in an instant. Which is not possible, of course.

Even though these comparisons afford some perspective, it is hard to relate to the given numbers, which basically refer to amounts of hundreds of molecules of gas in a million molecules of air (i.e. the current CO_2 concentration of 415 parts per million in the atmosphere). Not only are such gas concentrations hard to imagine for me as an earthbound being, I also lack a clear point of reference, which would tell me if such numbers are actually high or low. So, what about coming up with a more tangible measure of the impacts that my being and lifestyle have on my environment? Something more relatable down here on Earth's surface, where we all live. That was the starting point for Global Footprint Network and their concept of the *Ecological Footprint*, which is a much more comprehensive measure than just a carbon footprint. Here is what they mean by it: The **Ecological Footprint** is "[a] measure of how much area of biologically productive land and water an individual, population or activity requires to produce all the resources it consumes, to accomodate all its infrastructure, and to absorb the waste it generates, using prevailing technology and resource management practices. The Ecological Footprint is usually measured in global hectares. Because trade is global, an individual or country's Footprint includes land or sea from all over the world."[23]

Now, *that* I can relate to much better than to atmospheric concentrations of greenhouse gases. Although I don't yet know how large my EF is, I imagine it to be a few soccer fields worth of forest, cropland, grazing land for animals, a fishing pond, a bio-active waste dump, and so on. Whatever *bioenergy* the plants, animals, and microorganisms on that area of land and water produce, results in food I can eat, wood I can build with or burn, freshwater I can drink, and so on. But that bioenergy, or **biocapacity**, can also be converted into other forms of energy required to produce more resources I consume: toilet paper, a smartphone, air travel miles. Essentially *all* goods and services, as well as the energy required to produce them, are based on the *non-renewable resources* we extract from Earth (like minerals, metals, and fossil fuels) as well as the *renewable resources which Earth replenishes* each year (its *biocapacity*, in the form of biologically productive areas of land and water). Here is one concrete, comprehensible use of biocapacity: Approximately 50% of my Ecological Footprint is my carbon footprint (see piechart →), and I draw on forests to sequester and store away those

MY ECOLOGICAL FOOTPRINT

is about the size of this river island ↑ in the Amazon region. A little over 50% of it is forest, which I need to sequester and store my **carbon emissions**, represented by the red segment in the piechart →. Apart from that large carbon portion, the chart shows other types of land, which my EF is made up of: **cropland** where the food I consume is grown on, **fishing grounds** where the occasional fish I eat comes from, and so on. Of course, these types of land and water do not exist on a specific geographic patch of land all together. Rather, they are dispersed all around the globe.

Photo: iStock.com / filipefrazao. Graphic: Global Footprint Network

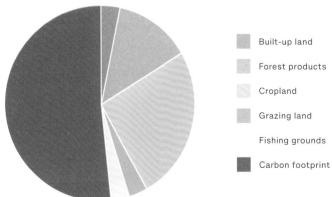

- Built-up land
- Forest products
- Cropland
- Grazing land
- Fishing grounds
- Carbon footprint

Rainforest, Rwanda
Biologically productive land and water is found almost everywhere on Earth. Including the animals, plants, funghi, protists, and microorganisms living on and in it, it makes up Earth's *biosphere*. But while the biosphere spans all of the planet, it is also a *finite* realm. Ever more humans with ever larger Ecological Footprints put ever more strain on that finite world. Photo: iStock.com / guenterguni

emissions (see p. 60). The trees, other plants, and microorganisms in the forest actually work so that a large portion of my carbon emissions will be safely stored away. The more emissions I cause, the more available biocapacity I will use exclusively for that purpose, and the larger my overall EF will be. Now we have an explicit measure for gauging whether my carbon footprint is "sustainable" or not: By taking into account that we live on a *finite* planet, which affords every earthling with only a proportionate share of as much biologically productive land and water as there actually is, we know where we stand. Currently, those biologically productive areas of land and water make up 12.2 billion hectares (equalling ca. 25% of the planet's total surface area).[24] Given the current world population of 7.8 billion people, consequently circa 1.6 global hectares are available per person (to keep things simple we disregard that other species also require part of that biocapacity).[25] Therefore, **an Ecological Footprint of 1.6 global hectares (gha) for every human being is compatible with the finite biocapacity of one planet Earth** (see **C** on p. 64).

Just as this accounting holds true for the whole planet, so it does for individual countries:[26] If we were to evenly distribute Austria's biocapacity to all its citizens (8.7 million), each would get 2.9 hectares. Fine, so we Austrians actually have almost twice the biocapacity that is required for a footprint compatible with one Earth! But here's the catch: the way they live, each Austrian requires not 2.9 but 6 hectares (!) of biologically productive land and water (see **A** on p. 64). For Americans, the ratio is worse: They each require 8.1 gha, but the USA provides them with only 3.6 gha. Even worse for a Qatari: their lifestyle requires 14.4 gha per capita, but their desert country only gives 1 gha. So how do the citizens of these high-income countries do it? Easy, they live on debt—they "borrow" the needed biocapacity elsewhere, and therefore become *biocapacity debitors*. They borrow it from Brazilians, for example, who require 2.8 gha to support their lifestyle, while their country—covered largely by the Amazon rainforest—provides them with such abundant biocapacity of 8.7 gha so that they can become *biocapacity creditors.*

But here is what this all boils down to for me: If I am anything like other Austrians, I require 3.5 times the amount of natural resources this planet can replenish each year. Yet, to keep my promise, stop global heating, and live sustainably from now on, I need to make ends meet with what *one* Earth can give!

Next time we meet (p. 102), we will begin to take stock of my consumption of energy, food, and so on, and see how I am doing. But given the kind of deficit I begin with, my hopes of achieving a "sustainable lifestyle" are dwindling even before we have started to take stock of anything . . . Can I still get out of this? I guess not because as I have just learned: No man is an island, entire of itself. So, here we all are, sitting in one big boat called *Earth*—and there actually is only *one* of her kind.

Causes

An oil refinery. Production sites like this exist all over the planet. We have one close to Vienna. You pass it when you commute between the city and its airport.

Photo: iStock.com / Suriyapong Thongsawang

Although I hardly ever see or smell them, **fossil fuels** coal, oil, and gas are around me all day long: We burn them inside the combustion engines of our vehicles, we heat our spaces and water with them, we cook with them. Besides its other applications in industry, the burning of fossil fuels also generates two thirds of electricity wordwide. This means we are all steeped in fossil fuels—not knee-deep but up to our necks.

THE GAS IS FLOWING AGAIN!

Burning fossil fuels for energy / Greenhouse gas emissions

Around half past eleven the plumber finally arrives, hooks up a diagnostic device to my gas boiler, and just as quickly as I got that fire going earlier, the boiler and the thermostat work again. The plumber is a man of few words, and just mumbles something about a firmware bug that shut down the thermostat. That shouldn't occur again, as he has now updated the system software. He makes me sign a form, and then is on his way again.

The gas boiler is now working properly again, drawing natural gas from the gas main, and burning it inside its burner. The heat from those flames flows into the radiators until the thermostat measures 20 °C room temperature and decreases the flow of gas to the burner. I let the fire in the wood stove die down, and we are back to business as usual with natural gas/methane (CH_4) heating the place once again. Of course what we are doing now is **burning a fossil fuel for heat energy**. That energy is produced the same way as before in the wood stove: The high-energy bonds between carbon atoms are broken up by the flames, and converted into heat. Divorced single carbon atoms partner up with two oxygen atoms into carbon dioxide, and that greenhouse gas exits my building via the flue, up into the air and on into the atmosphere.

But there is a major difference between the gas fire burning now, and the wood fire from earlier. That difference stems from where the burned carbon comes from—or rather, from *when* it comes. While the carbon in wood was stored by a tree a few decades ago, the carbon in fossil fuels is 300 to 360 million years old. It was deposited in the Earth during a geological period called the **Carboniferous**. "Fossil fuel use offers access to carbon stored from millions of years of photosynthesis: a massive energy subsidy from the deep past to modern society, upon which a great deal of our modern wealth depends."[27]

Here is what happened during the Carboniferous: When plants and other organic lifeforms (like micororganisms, algae, or animals) die but do not decay on land, and their remains instead sink into swamps, get deposited on the seabed as sediments, or end up underneath rock layers, they get pressurized in *anoxic* (devoid of oxygen) environments. Over long periods of time, that organic mass chemically transforms into other compounds, fossilizes or petrifies, and the carbon stored in it transforms into the **fossil fuels** we use for energy today: **coal, natural gas, and oil**. In contrast to wood, whose carbon was cycling in the natural carbon cycle (pp. 52–55) until recently, the carbon in fossil fuels was deposited on and in Earth toward the end of the Carboniferous, when a climate-related mass extinction of plants happened. It is these mass-graves that humans have been exhuming during the last couple of centuries.

In various locations around the globe, fossil fuels rose up to the surface. *Permafrost* soils contain vast amounts of frozen methane/natural gas (CH_4), surface coal fields in China have been mined for centuries, and in the now infamous Alberta *tar sands* (pp. 80–81) oil-soaked *bitumen* is skimmed from right beneath the removed topsoil. Most fossil fuels were buried deep underground though for hundreds of millions of years. It wasn't until great measures of them were dug up from the Earth during the nineteenth century oil rushes (see p. 116), and burned for the solar energy stored in them over millions of years, that vast amounts of hitherto tucked away carbon oxidized in air as CO_2 and became the main driver of our industrial civilization and of human-caused global heating.

But why does that pose a problem? Didn't we find out earlier that carbon always cycles between Earth's spheres via the natural carbon cycle? And doesn't that cycle regulate the temperature on Earth naturally? (pp. 52–55). Indeed, we did, and indeed, it does. But we also mentioned that this cycle takes up to millions of years to complete.[28] Nature is funny in that way: She follows strict rules—the natural laws—and even if something takes a thousand years to finish, or a million, or even a billion, you can be sure she will get the job done. The trouble may be though: If natural processes take so very long to complete, *you* might not be around anymore to see them taken across the finishing line.

And herein lies the problem with the massive amounts of carbon we have been pumping into the atmosphere since the beginning of the industrial era: We are not doing it in a measure of time anywhere near natural. Compare the 200 years we have been digging up and burning fossil fuels with the 300–360 million years it took for them to generate beneath Earth's surface! We are looking at two completely different timescales here (two of the clocks mentioned in the introduction). So, yes, the additional carbon we are putting into the atmosphere now so rapidly *will be* cycled down into the lithosphere through the weathering of rocks. That will even happen faster, as it gets warmer on Earth. But nature will still take millions of years to do so, not a few hundred.

Now that we have established *what* the problem is with burning vast amounts of fossil fuels, let's see *how big* a problem it is. You have surely come upon terms like "energy revolution" or "decarbonized economy." The urgency of achieving both is stressed regularly these days. But our global societal reality is still entirely another: **Roughly 86% of the world's primary energy demand is met by carbon-based fossil fuels coal, natural gas, and oil** (see graph p. 75 →).[29] While **renewable energy sources** like hydro, solar and wind power are on the rise (and their prices have dropped rapidly and below fossil fuels in recent years), their absolute share in total *primary energy supply* is still quite low (around 10%). *Primary energy* is the one contained in naturally occurring sources (like crude oil, sunlight, wind, or nuclear fission). For further use primary energy needs to be converted into *end-use energy* (like electricity, petrol, diesel, or biodiesel).

So, mankind still gets 86% of its energy from fossil fuels—despite the fact that they will be depleted at some point not far in the future.[30] New methods of getting at hard-to-reach deposits (like *fracking*, see pp. 88–89) are getting ever more risky and costly in recent years.

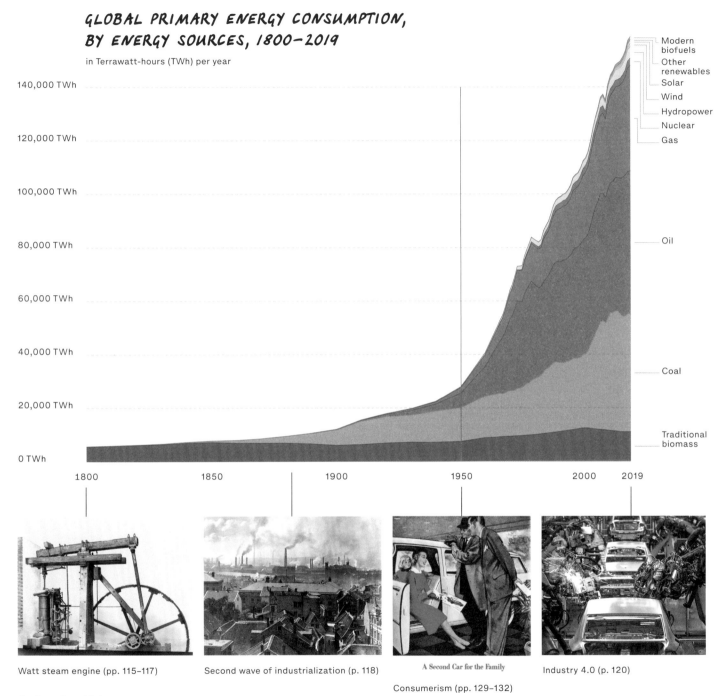

GLOBAL PRIMARY ENERGY CONSUMPTION, BY ENERGY SOURCES, 1800–2019

in Terrawatt-hours (TWh) per year

140,000 TWh

120,000 TWh

100,000 TWh

80,000 TWh

60,000 TWh

40,000 TWh

20,000 TWh

0 TWh

1800   1850   1900   1950   2000 2019

Modern biofuels
Other renewables
Solar
Wind
Hydropower
Nuclear
Gas

Oil

Coal

Traditional biomass

Watt steam engine (pp. 115–117)

Second wave of industrialization (p. 118)

A Second Car for the Family

Consumerism (pp. 129–132)

Industry 4.0 (p. 120)

Graphic top: Ourworldindata.org
Source: Vaclav Smil (2017) and
BP Statistical Review of World Energy;
Photos bottom: see referenced pages

86%

OF WORLDWIDE
PRIMARY ENERGY
DEMAND IS MET
BY FOSSIL FUELS
OIL, COAL, AND GAS.

← **Belchatów, Poland.** Belchatów Power Station is Europe's largest coal-fired power plant and also the continent's largest single emitter of carbon dioxide. The plant burns *lignite*—the dirtiest form of coal. It is operated by Polska Grupa Energetyczna (PGE). Photo: iStock.com / scyther5

↓ **The lights are on on planet Earth.** 64.5% of electricity worldwide is generated by fossil fuel-fired power plants. The image shows city lights in at least four US Gulf Coast states. Photo: NASA

↖ **Aruba, Lesser Antilles, Carribean.** The island's oil refinery (seen here in 2009) is located right next to beautiful beaches. While it may not be a pretty picture, it does not take us completely aback—that is how commonplace fossil fuels are in contemporary life.
Photo: Spencer Thomas (CC BY 2.0)

← **Los Angeles, USA.** Crossway junction at sunset hosting an endless procession of mostly combustion engine vehicles. It does not really bother us most of the time that we, keen on getting around in our cars, are burning the remains of plants and other organisms that have died over 300 million years ago.
Photo: iStock.com / franckreporter

In one year: 50.6 billion tonnes

OF GLOBAL GREENHOUSE GAS EMISSIONS*, BY SECTORS, 2010

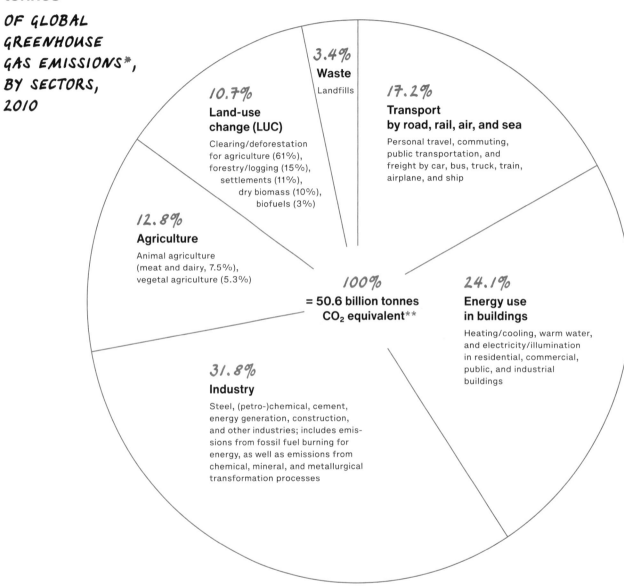

3.4%
Waste
Landfills

10.7%
Land-use change (LUC)
Clearing/deforestation for agriculture (61%), forestry/logging (15%), settlements (11%), dry biomass (10%), biofuels (3%)

17.2%
Transport by road, rail, air, and sea
Personal travel, commuting, public transportation, and freight by car, bus, truck, train, airplane, and ship

12.8%
Agriculture
Animal agriculture (meat and dairy, 7.5%), vegetal agriculture (5.3%)

100%
= 50.6 billion tonnes CO$_2$ equivalent**

24.1%
Energy use in buildings
Heating/cooling, warm water, and electricity/illumination in residential, commercial, public, and industrial buildings

31.8%
Industry
Steel, (petro-)chemical, cement, energy generation, construction, and other industries; includes emissions from fossil fuel burning for energy, as well as emissions from chemical, mineral, and metallurgical transformation processes

Graphic: SCHIENERL D/AD
Data source: BAJŽELJ/ALLWOOD/CULLEN (2013)

* GREENHOUSE GAS EMISSIONS

caused by human activity,
by gases

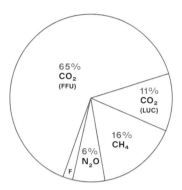

CO_2 Human-caused greenhouse gas emissions consist mostly of **carbon dioxide** (CO_2). Most of it stems from fossil fuel use (FFU, 65%), a smaller portion from land-use change (LUC, 11%).

CH_4 An additional 16% of GHG emissions are **methane** (CH_4, natural gas), most of which originates in animal agriculture *(enteric fermentation* in ruminants), waste disposal, direct energy use, and rice paddies. CH_4 is about 28 times more powerful a GHG than carbon dioxide, but luckily less persistent.

N_2O 6% of GHG emissions are **nitrous oxide** (N_2O), also originating mostly in agriculture. N_2O is even 256 times more powerful than carbon dioxide.

F gases A small portion of GHGs are **fluorinated gases** (F-gases): hydrofluorocarbons (HFC), perfluorocarbons (PFC), sulfur hexafluoride (SF_6).

** CO_2 equivalent (CO_2e)

is a unit of measurement for all greenhouse gases, which expresses the *global warming potential* (GWP) and persistency of a gas as compared to CO_2.

In addition, we have known for decades that burning fossil fuels results in **greenhouse gas (GHG) emissions** into the atmosphere: **65% of human-caused GHG emissions are carbon dioxide (CO_2) stemming from the burning of fossil fuels** (see piechart p. 79 ↙). An additional 11% CO_2 emissions stem from *land-use change* (LUC; see pp. 90–101). Combined CO_2 emissions from fossil fuel use (FFU) and LUC have totalled roughly 37 billion tonnes in 2019. Together, they make up 76% of man-made GHG emissions (in total, these amount to ca. 50 billion tonnes CO_2 equivalent per year). The rest is made up of other greenhouse gases, which are far more powerful, but luckily less persistent than CO_2. To ensure comparability nonetheless, **CO_2 equivalent (CO_2e)** has been introduced as a common unit of measurement for all greenhouse gases. It expresses the *global warming potential* (GWP) of GHGs as compared to CO_2, usually over a period of 100 years (GWP100). 16% of anthropogenic GHGs are **methane (CH_4)**—ca. 9 billion tonnes CO_2e in 2012, for example—, originating mostly in animal agriculture, waste disposal, direct energy use, and rice paddies. While methane is up to 28 times more powerful a GHG than CO_2, it is also far more volatile. 6% of GHG emissions are **nitrous oxide (N_2O)**, also originating mostly in agriculture through fertilizer use. N_2O is even 265 times more powerful than CO_2. 2% of anthropogenic GHG emissions are *fluorinated gases* (F-gases), which typically occur in coolants.

If you add all human-caused GHG emissions together and track them to their sources in various **societal sectors** (see piechart p. 78 ←), you find that **24% of GHGs stem from energy use in buildings (electricity, heating/cooling, warm water)** including residential, commercial, public, and industrial spaces. In that regard, it is important to note that ca. 64% of electricity worldwide is still generated by fossil-fuel-fired power plants (38% use coal, 23.2% gas, 3% oil).[31] Another 16% of GHG emissions stem from energy use (electricity/heat) in industrial processes. Including other emissions in industry from chemical, mineral, and metallurgical transformation processes, as well as from fossil fuel extraction itself, **industry causes ca. 32% of GHG emissions.** Another **23.5% originate in agriculture, related land-use change, and forestry** (see pp. 92–101). **17% stem from transport** (individual, public, and freight traffic by road, rail, air, and sea). The remaining **3.4%** are methane emitted from landfills. The numbers given here are based on data from a 2013 study by Bojana Bajželj and others[32] on world GHG emissions in 2010. While emissions have risen since then in absolute numbers, the proportions of sectors have remained largely unchanged.

If you run all those statistics by your inner accountant, you will find that the globalized world economy is running almost entirely on fossil fuels (and the consumption of other non-renewable resources). We all partake in that economy, whether we like it or not. Some benefits from the cheap and dirty energy source fossil fuels have trickled down to each and every one of us. To a large part it was their contribution that enabled the massive expansion of the world economy that

Alberta, Canada
At the infamous *Alberta Tar Sands* close to Fort McMurray oil sands are extracted directly from Earth's surface. *Tar* or *oil sands* are deposits of sand, clay, water, and *bitumen*, a viscous liquid that can be processed into oil products. Large areas of boreal forest had to be cleared and the land's top soil removed for this kind of operation. Photo: Orjan Ellingvag / Alamy Stock Photo

began with the industrial era (see pp. 114–121) and picked up pace considerably from the 1950s onward (see pp. 122–135).

But right now, for me, all these world statistics just boil down to facts like this one: Relatively cheap natural gas from Russia is heating my workplace, and I am thus emitting CO_2 every day and night during the winter. Combusting gas for heat energy is less carbon-intensive than burning oil or coal, but all these forms of generating heat emit greenhouse gases. So far, no viable new alternatives to heating with fossil fuels have presented themselves. One of the reasons why **renewable energy sources** have not gained more market share in primary energy supply is very simple but often overlooked and therefore even more perplexing: Renewables like solar, wind, and hydro excel at generating *electricity*—yet **electricity only makes up ca. 20% of end-use energy**. Almost all other end-use energy (for heating, transport, and industry) is generated by combusting fussil fuels, and it is much harder to substitute them with renewables in these sectors. We will delve into this in more detail in part *4. Solutions* of this book (pp. 186–205).

So while there are no options for me to substitute gas in heating thus far, I have opted to "reduce" my carbon footprint from using this fossil fuel. My utility provider offers "climate-neutral gas." That may seem like a contradiction in terms because how can burning gas be climate-neutral? But what they mean is that they **offset emissions** incurred by gas use by investing in carbon-neutral energy installations elsewhere. In my case that is a windpark and a hydropower dam in south-eastern Europe. This "offsetting" of emissions is regulated by the **EU emissions trading system (EU ETS)**, a principally sensible system but lacking scope and depth. The underlying mechanism is that the European Commission sets upper limits ("caps") on allowed GHG emissions for a number of heavily fossil fuel-dependent industries within its territory (primarily power stations, industrial plants, and inner-European aviation). The Commission issues *emission certificates* (allowances) to businesses in these industries, which are valid for a certain period. If a company emits *fewer* GHGs than allowed for by their certificates, it can sell these at a fixed price to another business emitting *more* than their allowance ("trade"). In this *cap-and-trade system* overall allowed emissions are gradually reduced over time, and the fixed price for carbon emissions is simultaneously raised. This is supposed to create a strong and predictable incentive for businesses to lower their carbon footprint over time. The two basic problems with the EU ETS are: 1. It does not include all industries and societal sectors (scope). EU-wide the ETS covers only 45% of overall GHG emissions (in Austria, the ratio is at an even lower 36%). 2. The pricing of carbon emissions is still very low (24 Euros per tonne CO_2), plus: a considerable portion of certificates is issued for free. If those two problems were overcome, such a cap-and-trade system—perhaps implemented worldwide with a unitary price for carbon emissions—could become a major factor in solving the climate crisis (see pp. 204–205).

Top 10

CLIMATE-POLLUTING COUNTRIES, 2017

Total greenhouse gas emissions in billion tonnes CO_2 equivalent per year

1. China (13.119)
2. USA (6.457)
3. EU-28 (4.317)
4. India (2.940)
5. Russian Federation (2.640)
6. Japan (1.290)
7. Brazil (1.035)
8. Indonesia (0.895)
9. Canada (0.716)
10. South Korea (0.700)

Data source: https://climateactiontracker.org/countries/

GLOBAL CO₂ EMISSIONS FROM FOSSIL SOURCES, 1800–2018

in billion tonnes per year

Global carbon dioxide emissions from fossil sources coal, oil, gas, cement production (8% of total), and *flaring* (at many oil and gas production sites around the world, superfluous natural gas, which might over-pressure plant equipment, is simply and wastefully burnt off to release pressure).

Mind the sharp exponential increase of emissions from 1950 onward: they have soared by 750% (!) in the last 70 years alone. This period has therefore become known as the *Great Acceleration* (see pp. 122–135).

(CO₂ emissions from land-use change are not separately shown in this graphic.)

Graphic: Ourworldindata.org.
Data source: Global Carbon Project (GCP) and Carbon Dioxide Analysis Center (CDIAC)

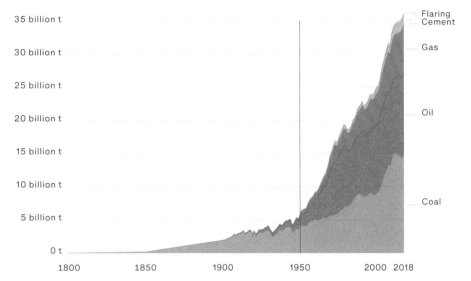

35 billion t
30 billion t
25 billion t
20 billion t
15 billion t
10 billion t
5 billion t
0 t

1800   1850   1900   1950   2000 2018

Flaring
Cement
Gas
Oil
Coal

Surface coal mine in Shuo Zhou City, Shanxi, China, 2004. The miners are preparing to blast the coal. Photo: Peter Essick / Cavan Images

Indian women sorting and collecting coal and anthracite at a surface coal mine.
Photo: Hakbong Kwon / Alamy Stock Photo

GLOBAL CO₂ EMISSIONS, BY WORLD REGIONS, 1751–2018

Annual carbon dioxide emissions attributed to world regions

While emissions peaked in the **EU countries** already in 1979, they stagnated throughout the 1980s, 1990s, and 2000s. Since the financial crisis of 2007/2008 they have been decreasing slightly. In the **US** peak emissions were reached only in 2007 just before the financial crisis. Since then emissions have shown no clear upward or downward trend but remain on a high level. In **China** emissions have risen enormously since the early 1980s, not least because the country has become the Western world's "workbench" since then (1980: 1.49 billion tonnes CO_2; 2018: 10.06). **India** and the **Middle East** (part of *Asia [excl. China and India]*) show continued strong growth in emissions since the 1980s.

Graphic: Ourworldindata.org.
Data source: Carbon Dioxide Analysis Center (CDIAC) and Global Carbon Project (GCP)

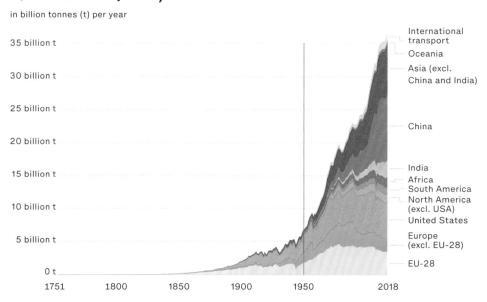

in billion tonnes (t) per year

International transport
Oceania
Asia (excl. China and India)
China
India
Africa
South America
North America (excl. USA)
United States
Europe (excl. EU-28)
EU-28

Top 10

CLIMATE-POLLUTING COUNTRIES, BY PER CAPITA CO₂ EMISSIONS, 2017

in tonnes CO_2 per capita per year

1. Qatar (30.4)
2. Kuwait (21.6)
3. United Arab Emirates (20.9)
4. Bahrain (20)
5. Saudi Arabia (16.2)
6. Australia (15.6) | Brunei (15.6)
7. Canada (15)
8. USA (14.6)
9. Kazakhstan (14.2)
10. Trinidad and Tobago (13.2)

Mid-table

Japan (8.9) | Singapore (8.5)
South Africa (7.4) | Israel (7.3)
China (6.7) | New Zealand (6.7)
Iceland (6.3) | EU-28 (6.3)

Bottom of the pile

Ethiopia (0.13) | Niger (0.12)
Mali (0.09) | Malawi (0.07)
Central African Republic (0.07)
Democratic Republic of
the Congo (0.06) | Chad (0.05)
Somalia (0.05) | Burundi (0.05)

High per capita emissions in the Top 10 countries do not necessarily indicate excessive fossil fuel use by individuals. They rather stem from high levels of fossil fuel extraction in those countries, which cause high CO_2 emissions. Often though the two factors *do* coincide.

Data source for top and mid 10: https://www.iea.org/data-and-statistics, Energy topic > CO₂ emissions, Indicator > CO₂ emissions per capita: Data source for bottom 10: https://ourworldindata.org/per-capita-co2

Besides the offsetting-option I chose for my office, I am lucky that my apartment receives *district heating:* As in other larger cities, in Vienna this eco-friendlier heating system supplies a third of all buildings with thermal energy. It works by heating water or steam in a centralized location and then distributing them via a pipe network to residential buildings. The energy needed to heat the steam or water is delivered by coupled heat/power installations and by industrial waste heat (two thirds) plus heat generated in waste incineration plants (one third). This process results in 70% fewer GHG emissions per delivered energy unit than heating with natural gas.

As for my personal electricity use, I am also lucky to be living in Austria: The country is geographically and topographically well-endowed with opportunities for **renewable energy installations**. Large-scale **windparks** are located in the flat and windy North and East, while the western and southern alpine regions allow for large **hydropower dams**. Already, the country is producing 70% of its electricity from renewable sources. New policies aim to increase that ratio to 100% by 2030. But while Austria is lucky with its landscapes, there may also be a huge downside looming in the future: with alpine glaciers melting away rapidly in the changing climate, the hydrological cycle in Austria could also change substantially. Highly productive hydropower might then get into trouble. In addition to wind, hydro, and solar, more and more **biomass power plants** are being installed in the countryside of Austria, which is rich in forest land. These power plants burn the remains of trees and other plants, as well as biowaste to generate heat, which is supplied to localized clusters of households.

But even if some countries like Austria are lucky and a little ahead with that "energy revolution," we as Earth citizens collectively face a dilemma now: Science tells us unequivocally that we have to completely *decarbonize* our economies within the next twenty years so as to avoid life threatening global heating beyond 2 °C. Sounds impossible? Maybe. But whatever brought us to this absurd juncture is entirely of our own making. Therefore we are also the only ones who can and have to get ourselves out of this quagmire. We'd better hurry though, because I promised I would stop global heating by tonight. Also, I am getting hungry now . . .

Top 10

CLIMATE-POLLUTING COUNTRIES, HISTORICALLY

Cumulative CO$_2$ emissions
from fossil fuel use, 1751–2017,
by country, in billion tonnes
(% of total of 1,575 billion tonnes)

The climate crisis is characterized by rising global
temperatures, which are driven by an amplified
greenhouse effect. That amplification is due to
ever more human-emitted greenhouse gases
accumulating in the atmosphere. Since CO$_2$ in
particular lingers there for up to millenia, one
question is of great importance:
**Which countries have historically contributed
most to the climate crisis?**
The United States are by far the most climate-
polluting country historically (399 billion tonnes
of CO$_2$, or 25% of cumulative emissions). In
second place come the countries of the EU-28
(353 billion tonnes, 22%), in third—but by a large
gap—China (200 billion tonnes, 12.7%), followed
by the Russian Federation (6%), Japan (4%),
India (3%), Canada (2%), and South Korea (1%).
The question of historical responsibility for the
climate crisis will play a major role in our discus-
sion of climate justice and equity (pp. 177–181).

Graphic: ourworldindata.org
Data source: Global Carbon Project (GCP)
and Carbon Dioxide Analysis Center (CDIAC).

For more individual country statistics see:
https://ourworldindata.org/uloads/2019/
10/Cumulative-CO2-treemap.png

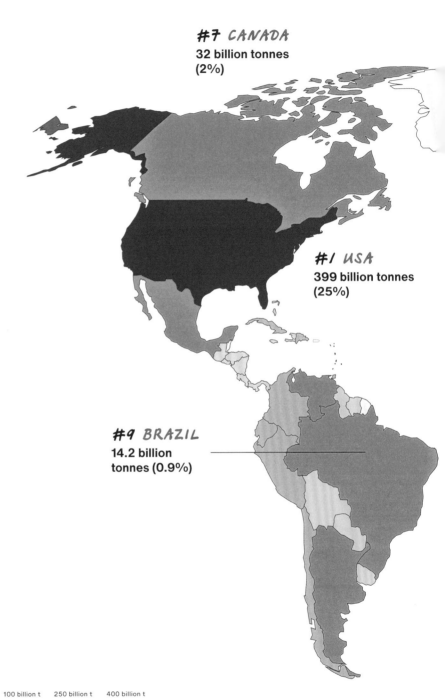

#7 CANADA
**32 billion tonnes
(2%)**

#1 USA
**399 billion tonnes
(25%)**

#9 BRAZIL
**14.2 billion
tonnes (0.9%)**

No data 0 t   50 million t   500 million t   5 billion t   50 billion t   100 billion t   250 billion t   400 billion t

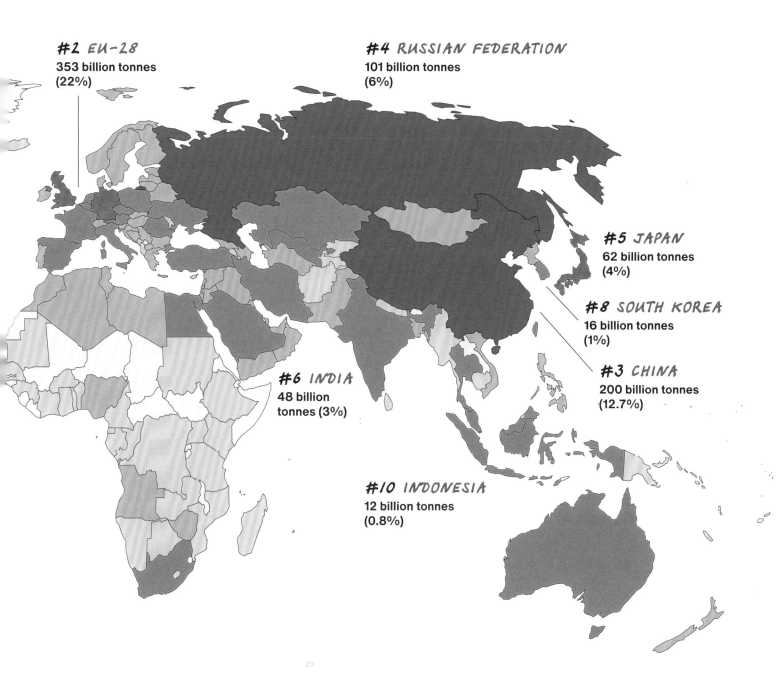

#2 *EU-28*
353 billion tonnes
(22%)

#4 *RUSSIAN FEDERATION*
101 billion tonnes
(6%)

#5 *JAPAN*
62 billion tonnes
(4%)

#8 *SOUTH KOREA*
16 billion tonnes
(1%)

#3 *CHINA*
200 billion tonnes
(12.7%)

#6 *INDIA*
48 billion
tonnes (3%)

#10 *INDONESIA*
12 billion tonnes
(0.8%)

Top 20

CLIMATE-POLLUTING FOSSIL FUEL COMPANIES, 1965–2017

Cumulative greenhouse gas emissions, 1965–2017, in billion tonnes CO_2 equivalent (total for Top 20: 480 billion tonnes = 35% of all energy-related emissions worldwide)

state-owned / investor-owned

1. Saudi Aramco (59.26) Saudi Arabia
2. Chevron (43.35) USA
3. Gazprom (43.23) Russia
4. ExxonMobil (41.90) USA
5. National Iranian Oil Co. (35.66) Iran
6. BP (34.02) UK
7. Royal Dutch Shell (31.95) The Netherlands
8. Coal India (23.12) India
9. Pemex (22.65) Mexico
10. Petróleos de Venezuela (PDVSA) (15.75) Venezuela
11. PetroChina/China Natl Petroleum (15.63) China
12. Peabody Energy (15.39) USA
13. ConocoPhillips (15.23) USA
14. Abu Dhabi (13.84) United Arab Emirates
15. Kuwait Petroleum Corp. (13.48) Kuwait
16. Iraq National Oil Co. (12.60) Iraq
17. Total SA (12.35) France
18. Sonatrach (12.30) Algeria
19. BHP Billiton (9.80) Australia
20. Petrobras (8.68) Brazil

Data source: Climate Accountability Institute (CAI). *Carbon Majors. Update 8 October 2019*, retrieved Dec. 28, 2019 from: http://climateaccountability.org/carbonmajors.html; also see CDP CARBON MAJORS REPORT (2017)

Jonah Field, Wyoming, USA
Jonah Field is a large natural gas field in
the Green River Basin. The photo shows
its scarred landscape dotted with *hydrau-
lic fracturing* ("fracking") infrastructure.
Fracking is used to fracture bedrock
formations by injecting highly pressurized
liquids. Otherwise inaccessible natural
gas or oil deposits can thus be accessed
through the resulting cracks. The tech-
nique is highly controversial as it leads to
groundwater and surface water contami-
nation, methane leakage, and increased
seismic activity.
Photo: Bruce Gordon – Ecoflight (CC BY 2.0)

Our diets play a critical role in linking human health and environmental sustainability. Only if that link is firmly established will the biosphere not be overburdened with feeding 10 billion people in 2050.

Photo: iStock.com / filadendron

LUNCH BREAK

Food, agriculture, and the climate

I go out for lunch on a lot of workdays, mostly because I am not that great a cook, and I don't like heating up processed food. Sometimes Nina joins me for lunch, as is the case today. We go to the small place just around the corner, where they cook deliciously, and use organically farmed regional produce and meat. Yes, meat. I do still eat fish and chicken sometimes: I am a *flexitarian*, as they say, meaning that my diet is largely plant-based but includes modest amounts of fish, meat, and dairy foods. I do not eat red meat anymore though (beef, pork, lamb) for a number of reasons, which I will get into later.

Today, I go for a classic at this place: the *Waldviertler Seitan-Schnitzel*, served with potatoes and roasted young spinach. This dish is a deserved classic for a number of reasons: 1. It tastes deliciously. 2. While being a vegetarian meal, it tastes and feels almost exactly like a *Wiener Schnitzel* (a classic meat-based Viennese dish). 3. It is made of *seitan*, a starchless wheat-protein, produced organically close to Vienna. The latter reason makes this dish a healthy food for me but also a perfect ingredient for a "planetary health diet."

"Planetary health diet? What's that?" Nina asks. Obviously, I just said that last part out loud—which I was not aware of . . . but hey, my blood sugar is low, and I smell all these delicious dishes swirling around me. So, I try to muster what strength remains and explain the term **planetary health diet** by entering it in my favorite search engine Ecosia, and citing the scientists—Walter Willett and Johan Rockström—who coined it. The term is supposed "to highlight the critical role that diets play in linking human health and environmental sustainability and the need to integrate these often-separate agendas into a common global agenda for food system transformation to achieve the . . . Paris Agreement."[33] Okay, with that definition we have a lot on our plate to digest (intellectually). So let's first take a few steps back and look at the big picture in terms of food, agriculture, and the climate.

Right now, 7.8 billion people inhabit Earth. That number is up from 1.65 billions around 1900 and will probably grow to almost ten billions by 2050.[34] So, between 1900 and 2020, world population has increased by six billion people! You can very well speak of a **population explosion**. Which is great, don't get me wrong: Many more people began living longer and healthier lives during the twentieth century. Improvements in life expectancy and healthcare came from two important inventions/innovations early during the century. The first one is the antibiotic *Penicillin*, first described by Alexander Fleming at the end of the 1920s, and then developed into antibiotic medication during the Second World War. It greatly improved human health prospects: Life expectancy doubled from 35 to 70 years globally within just the twentieth century. The second important innovation was the "Haber-Bosch process," which not everyone might have heard of, although it has benefited mankind equally if not more than Penicillin. The **Haber-Bosch synthesis** (named after German chemist Fritz Haber and industrialist Carl

Bosch) allowed for large quantities of chemical fertilizer to be produced "out of thin air," so to speak. In an energy-intense process (enabled by fossil fuel use), *ammonia* can be made from freely and abundantly available nitrogen in Earth's atmosphere (78% share, see p. 43). Application of ammonia as the active ingredient in **chemical fertilizer** greatly increased crop yields all over the world, as it shortened crop cycles—waiting for natural soil nutrients to regenerate became obsolete.[35] This also allowed for large-scale **industrial agriculture** to be established, which could feed ever more people.[36]

Global food production of calories has generally kept pace with the "population explosion" over the course of the twentieth century, which was no easy feat to achieve. The preferred operating mode of the agricultural industry would become **large-scale monocultures**—single kinds of crop cultivated on large contiguous pads. They can be worked by mechanized planting and farming equipment, and yields per hectare can be boosted with **chemical fertilizers, pesticides, and herbicides** (like glyphosat).

But even more resources have to go into that mode of production: **agriculture uses 70% of Earth's freshwater supply** today, and it has led to the **conversion and use of 40% of Earth's land area for farming**.[37] Considering how much of the planet is actually suited as arable land, those 40% are already the maximum of what can be achieved—unless you clear ever more forests for that purpose. Which is exactly what has and is still happening, and why conversion into agricultural lands has also become the most important factor in driving **biodiversity loss**.[38] More and more natural ecosystems are "ingested" into the food production system, and thus annihilated. You can see the workings of this system in nightmarish excess, with **slash-and-burn land clearance** running amuck in Brazil (see p. 163). Large areas of formerly highly biodiverse virgin rainforest—habitat to indigenous peoples and countless species of birds, insects, and vertebrates—have been turned into **grazing land for cattle, but mostly areas under cultivation for feed plants like soy. Palm oil production** (and extraction of timber, rubber, and minerals) drives rapid decimation of enormous areas of virgin rainforest on the Indonesian islands of Sumatra and Borneo. Between 1980 and 2000 alone, globally, 100 million hectares of tropical forest were lost to cattle ranching, plantations, and such.[39] That is the size of a country like Ethiopia. All these changes to and destruction of natural ecosystems are subsumed under the rather technical term **land-use change (LUC)**. Historically, deforestation and other land-use change has caused one third of the total ca. 2,400 billion tonnes of CO_2 emitted by human activity since the beginning of the industrial era.[40] Today, that portion is down to ca. 17% of annual CO_2 emissions (or 11% of overall greenhouse gas emissions, see pp. 78–79), but that amount still equals huge areas of forest being cleared (most often burnt down) every year to make way for agricultural land. When that happens, the carbon stored in these forests/trees gets released back

→

Global food production

HAS BECOME A MAJOR DRIVER OF ENVIRONMENTAL CRISES. AT THE SAME TIME, IT IS ITSELF SEVERELY IMPACTED BY THOSE CRISES.

→ **Increasingly frequent and intense extreme weather events** (heatwaves, droughts, deluges, floods, storms) disrupt crop cycles and diminish agricultural yield. The image shows a bone-dry field in Germany during the heatwave of 2018.
Photo: iStock.com / ollo

↘ **Feeding a world population of 10 billion people in 2050** will be a major global challenge. View of a crosswalk in Mexico City, Mexico.
Photo: iStock.com / Orbon Alija

↓ **Over-use of land and freshwater** threatens biodiversity (as more and more natural ecosystems are "ingested" into the agrobusiness) and disturbs the natural water cycle on Earth (agriculture uses 70% of worldwide freshwater). The image shows typically patchworked agricultural land in Austria. Photo: Christian Schienerl

↓ **Over-use of chemical fertilizers, herbicides, and pesticides** has led to serious soil degradation in many parts of the world. These practices are also altering the natural stocks and flows of nitrogen and phosphorus around the globe (see pages 98 and 152). The image shows a self-propelled sprayer applying herbicide on a field of young corn in Vinnitsa, Ukraine, 2018.
Photo: iStock.com / Oleksandr Yuchynskyi

DEFORESTATION:
PERUVIAN AMAZON IN 1986 . . .

Near Pucallpa, Peruvian Amazon
Nov. 13, 1986 – Oct. 30, 2016

Large areas within the Amazon rainforest
have undergone deforestation over the
past few decades. However, in locations
like the Peruvian Amazon, most of the
deforestation has been caused by small-
scale agriculture in recent years, according to the Monitoring of the Andean
Amazon Project. These images show land
ca. 40 kilometers northwest of Pucallpa
along the Aguaytia River. Lush green
dominates the 1986 image (left page),
while deforested land is light green or
pink in the 2016 image (right page). Two
large-scale oil palm plantations dominate
the 2016 image.

Images taken by Landsat. Source: U.S. Geological
Survey (USGS) Landsat Missions Gallery: "Monitoring
Deforestation in the Amazon"; U.S. Department of the
Interior / USGS and NASA

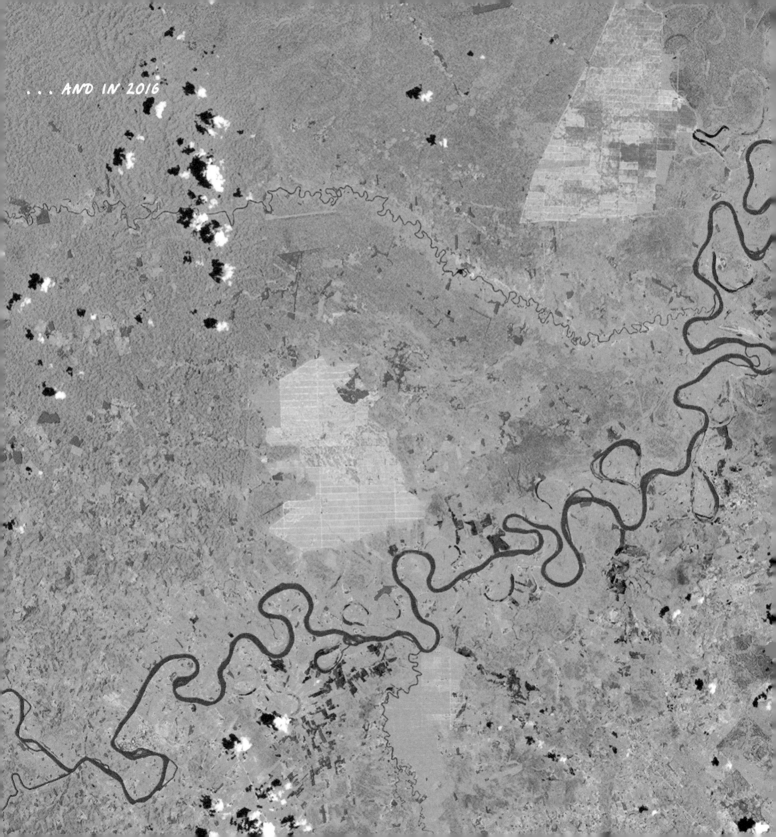
... AND IN 2016

↑ **The Amazon, Northern Brazil.**
Brazil is currently the seventh largest climate polluter in the world. Most of its GHG emissions stem from land-use change and deforestation (as shown in the picture). The vast majority of cleared land is used for growing feed for livestock, for cattle ranching, and for oil palm plantations.
Photo: iStock.com / **luoman**

→ **Factory farming.**
Cattle at the farm "Estância Bahia" in Brazil. Animal agriculture and related land-use change account for 14 to 18% of global greenhouse gas emissions.
Photo: Daniel Beltrá / Greenpeace

Today, 97%

OF LAND ANIMAL BODY MASS ARE MADE UP OF HUMANS (30%) AND THE LIVESTOCK THEY RAISE (67%) . . .

to the atmosphere as CO_2. Moreover, the biosphere's capacity to act as a *carbon sink* is thus decimated (see pp. 58–63).

All these environmental degradations brought about by industrial agriculture tipped over that once marvelous process of feeding billions of people into something highly unsustainable. But it is not *feeding that many people* that constitutes the problem—actually 80% of the world's food in nutritional value terms are produced by family farmers[41]—rather it is *what* people are being fed by the agriculture industry. One type of industrial food stands out in terms of how resource-intensive it is: **red meat and dairy products**.

25% of global greenhouse gas emissions stem from agriculture and related land-use change. **Animal agriculture and directly associated land-use change alone account for 12% of global GHG emissions** (see p. 78). Yet a large share of newly deforested land worldwide is not used to grow things we humans can eat. Instead it serves for growing feed (mostly grain and soy) for the animals we humans eat. If you believe this to be a marginal problem, be prepared for your world to be turned upside down: **Only 3% of all terrestrial vertebrate animals are still wildlife**. 97% (!) of land animal bodymass are made up of humans (30%) and the livestock they raise (67%).[42] Worldwide, **77% of agricultural land is used as grazing grounds and for growing feed for animals**.[43] In addition to GHG emissions associated with land-use change for animal agriculture, especially **cattle farming** is a major contributor to global heating in yet another way. As a by-product of their digestive process (**enteric fermentation**), cows and other ruminants release **methane (CH_4)** into the atmosphere. With methane being up to 28 times more powerful a greenhouse gas than CO_2, the combined GHG emissions associated with producing 1 kg of beef are 12 times higher than those for 1 kg of vegetables.[44] If you add all other resources required to raise livestock to the equation—14 times more land than for growing vegetables, almost 200 times more water, a globalized system of production and shipment of feed—**production of red meat is mind-bogglingly wasteful and polluting**. While consumption of red meat is somewhat declining in high-income countries, the fastest growing regions of the world (mainly China and India), also develop a growing hunger for meat.

Beyond meat production there are even more sustainability issues that today's global food production system is fraught with:
▪ **Crop yields** in Europe, Northern America, and parts of Asia—the regions where chemical fertilizers have been used most intensely over the past decades—are decreasing. Soils in these world regions have been seriously degraded. Consequently, more and more fertilizer has to be used to maintain yield levels. Proposals have been made to **distribute remaining stocks of fertilizer free of charge to other world regions**, which promise higher yield efficiency (e.g. in Africa). Crop yield decreases are compounded by increasingly frequent, intense, and long-lasting **extreme weather events**.

- Excessive use of chemical fertilizer also leads to ever more nitrogen (N) and phosphorus (P) running off from soils and entering freshwater and then coastal ecosystems (also see pp. 152–153). This has produced "more than 400 hypoxic zones [= coastal dead zones] that affected a total area of more than 245,000 km² as early as 2008."[45] Not only do coastal dead zones constitute environmental disasters per se, they also endanger food security from fisheries all around the world: 90% of life in the oceans is found in the shallow seas close to coasts.[46] Due to industrial overfishing for decades, many fishstocks all around the globe have been depleted to dangerous levels as it is.[47] Now coastal dead zones exacerbate this problem.

- According to the Food and Agriculture Organization of the United Nations (FAO) about one third of food produced worldwide never makes it onto anybody's plate.[48] Along the whole supply chain from harvest, via processing to wholesale food losses occur, whereas in retail and finally in people's homes a lot of food is wasted. These losses and waste pose a serious problem for global food security.

- One billion people worldwide suffer from obesity (over-weight with serious associated health risks), while one billion are under- or malnourished.

All these mutually reinforcing problems illustrate the urgent need for a radical rethinking and revolution of the agricultural sector. Establishment of a science-based, globally thought-out and locally enacted *(glocalized)* system of food production is called for, which simultaneously considers human health as well as the planet's health (for more on the "planetary health diet," see from page 201).

I pause because right now our food arrives. Nina and I take a few bites in silence, appreciative of the food on our plates: besides having enough of it, we also know that it was produced sustainably, organically farmed in the region where it is consumed. But sustainably feeding 10 billion people in 2050 will be a whole other challenge. Besides important efforts for chemical-free intensification of crop yields (see p. 202), the only other real game-changer is a drastic reduction of animal-source food and its replacement with plant-based alternatives.[49] Actually, this is good news: we will eat healthier, live longer, mistreat and kill fewer animals, and be able to sustain ourselves on the one finite planet we find ourselves on.

THE REMAINING 3% wildlife HAVE BEEN PUSHED TO THE BRINK OF EXTINCTION.

Deforestation in Central Kalimantan, Indonesia.
A network of access roads on former orangutan habitat inside the PT Karya Makmur Abadi Estate II palm oil concession.
Photo: Ulet Ifansasti / Greenpeace

Orangutan in Central Kalimantan, Indonesia.
Baby orangutans at the Orangutan Foundation International Care Center in Pangkalan Bun, Central Kalimantan. Expansion of oil palm plantations is destroying their rainforest habitat. Photo: Ulet Ifansasti / Greenpeace

77 %
OF AGRICULTURAL LAND WORLDWIDE ARE USED JUST FOR raising livestock and growing feed.

Deforested area in the Amazon, Brazil
Large-scale fields of soy, which is grown as feed for livestock all over the world, supplant highly biodiverse virgin rainforest. Greenpeace documents a number of areas in the Amazon, looking at people, natural wildlife and the impact that the soy industry is having on the landscape. Photo: Daniel Beltrá / Greenpeace

MY FOOTPRINT 2

– ENERGY
– FOOD

It is past midday—time to check how I am doing with stopping global heating: While we looked at what an Ecological Footprint (EF) actually is before (pp. 64–69), I will now take stock of some of the resource uses which make up my personal EF.

Before we begin though, let me clarify that we are looking at the Ecological Footprint I leave as an individual. We are not considering the footprint that my work or services as a graphic designer cause (who has his own office and company car, uses electricity, computers, other working materials, and so forth). The resources used and embodied in my work and in the services I provide to other people, businesses, and organizations are not part of *my* footprint but will instead show up as consumption of "goods" in the Ecological Footprints of my clients. We will find out how that works exactly in the next footprint section (p. 136). For now let's just say that footprint accounting is focussed on the consumption side of things, not the production side. The vegetables I eat and the haircut I get are part of *my* footprint and not of those of the farmer who grew the vegetables or the barber who gives me a haircut.

In this section though, I will begin by taking stock of other things anyhow: my **household energy use for heating, warm water, and electricity** and my **food consumption**.

I have answered all kinds of questions regarding my household and the kind of energy I use in it on footprintcalculator.org. Now the results of the calculation show that energy use is responsible for a 17%-portion of my EF—which is not that bad, I guess; it is actually less than I would have expected. But where does my **energy use at home** land me in relation to the Austrian average [A] and the world average [B]? As you can gauge from the size of the circular segments to the right, I manage to stay quite far below A, and just slightly above B. This is due to several factors. First, 100% of my electricity at home comes from renewable energy sources, as I have switched to an energy provider offering that option. Austria is lucky in that regard, as currently 70% of its electricity is produced by hydropower dams in the Alpine region and along its rivers and by windpower parks in the country's windy East. Current government policies even aim to raise that quota to 100% electricity from renewable sources by 2030. Second, my heating at home is provided as *district heating*, which results in 70% fewer greenhouse gas emissions than heating with natural gas (see p. 85). In addition, I share my 90 m² apartment with a flatmate—so my space requirements are far from excessive.

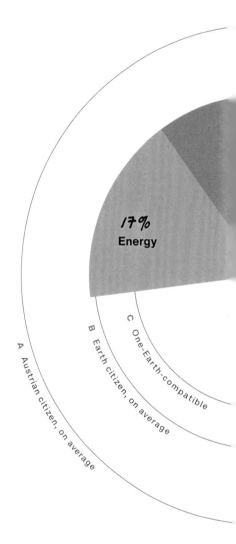

17%
Energy

A Austrian citizen, on average

B Earth citizen, on average

C One-Earth-compatible

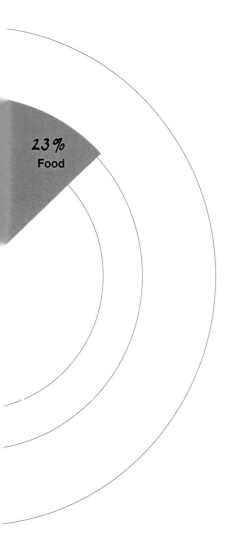

23%
Food

In terms of **food consumption**, which is responsible for a 23%-portion of my EF, I also stay well below the Austrian average [A], and just slightly above world average [B]. While that portion is larger in relative terms than I would have thought, its absolute size again is quite small in comparison to A. The primary reason for that may be that I eat hardly any meat (except for chicken and fish occasionally) and few eggs and dairy products (cheese from time to time). In addition, I buy *regionally sourced*, *organically farmed*, and *unpackaged* food whenever possible. Also, I try to stay as *seasonal* as I can, meaning that I eat produce (vegetables and fruit) at the time of the year when it is harvested in Austria or buy vegetables and fruit which can be stored for longer periods of time (apples or potatoes, for example). On the other hand, I do eat some processed food, I eat out quite a lot, and I am an addict as far as coffee and chocolate are concerned. The latter are neither locally produced nor unprocessed foods—but at least I buy them fairly traded. So, a 23%-portion of my overall footprint for food consumption seems justified. I am glad I have stopped eating red meat (beef, pork, lamb) though, else my food segment would have been much larger still, both relatively and absolutely.

As we have seen, for both energy use and food consumption I manage to stay well below the Austrian average [A] and just slightly above World average [B]. But I am still way above [C], a footprint compatible with the natural resources of one Earth. That means, so far I am failing to stop global heating by tonight. What a bummer!

Nevertheless, I am genuinely surprised that a few modest lifestyle changes—eating hardly any meat, buying local, organic, and seasonal food, switching to 100% renewable electricity—have taken me that far already. Right now, my Ecological Footprint is half that of the Austrian citizen, on average. No small feat, I would say! In addition, I should stress that none of the modest lifestyle changes I have made make me feel like I am missing out on something or had to relinquish something. On the contrary: I feel fitter for eating more healthily and better about myself for living less wastefully. But there is still huge progress to be made if I want to get to an EF compatible with one Earth. Hopefully, I will do better with the other portions of my footprint, which are still coming up (pp. 136, 170). But I feel energized and inspired now to go all the way!

For further tips on how to reduce your energy use and the emission intensity of your food consumption, see part **4. Solutions** *on pages* 189, 193 *and* 201.

Cruising along the highway, sitting in my petrol engine vehicle (PEV) alone— as do most other drivers.
Photo: Christian Schienerl

DRIVING TO THE PRINTERS

Mobility, "externalities," and carbon pricing

It's half past one—I have to get going and head out to the printers who are located about 35 kilometers outside of Vienna. This afternoon, we will print the book that I sent final corrections for in the morning. I will check the printing of a few dozen pages. If everything runs smoothly, I should be on my way back to the office within two hours.

For convenience, time-saving, and often because I have heavy (paper) stuff to carry around, I take the car to get to and back from the printers. I drive a small petrol engine vehicle (PEV), which I bought used five years ago. Per year, I do approximately 6,000 km on it—3,000 km for business purposes, the rest for personal purposes. This ride to and back from the printers will be 70 km, mostly along the highway, which I presently get on to. As usual around this time of day, traffic is heavy but flowing. I should get to my destination within 20–25 minutes.

With some time to spare, I try to put some things into perspective regarding my mobility. So, I am sitting in a combustion engine vehicle alone, which—under present driving conditions—guzzles ca. seven liters of petrol per 100 km. This 70-km-ride to and from the printers will therefore consume five liters of petrol. With current petrol rates (at 1.25 Euros/liter), this ride will therefore cost 6.25 Euros for fuel. For comparison: Public transportation for the same itinerary (a commuter train) would cost 9.60 Euros. That is quite a lot more. But wait! There is something wrong with my comparison. By comparing costs that way I have already made a gross miscalculation: I have not included all other costs, such as maintaining my car, insurance, etc. to the tally for the car ride. So I divide all these costs (roughly 3,000 Euros per year) by the kilometers I do in a year (6,000), and multiply the resulting 50 Cents per kilometer with the 70 km for this ride. That calculation adds 35 Euros to the bill! Oops, that's a lot more than I would have thought. Now admitted, I do get back about half of that additional cost through tax returns. But even when I account for that, the true monetary cost of this 70 km ride ends up being almost 24 Euros. Quite a bit more than the train ride, which now doesn't seem expensive anymore at all.

As long as you consider the price per *passenger kilometer* travelled—that is, moving a single passenger a distance of one kilometer—the train is about 60% cheaper than the car. But if you consider that the train carries many more passengers—even if its operational and maintenance costs are much higher than for a car—the train will end up being much more expensive. Conversely, you could argue that individual transport on a petrol engine car is far too cheap . . . While you can and should make a case for both sides, there are other costs—hidden from plain view—which will alter the comparison yet again.

Right now, these costs do not enter the equation because in standard economics they are considered so-called "externalities." **Externalities** are impacts of economic decisions incurring costs that no one pays for and no one is compensated for. Typically, environmental pollution stemming from the manufacture and use of products was considered an externality in the past, because the environment did not present us with an invoice for polluting and degrading it. But this is now changing, as awareness for the deficiency of such reasoning grows—but also because nature now indeed does present us with such invoices. Just think of the growing financial damages due to more frequent and intense extreme weather events. As for the case of my car ride to and back from the printers, the most crucial "externality" is this: Among other pollutants, my small car directly emits ca. 130 grams of CO_2 per kilometer driven (for the purpose of the following calculations, I will only consider *direct emissions* stemming from the car's operation but not *indirect emissions* associated with the car's production, recycling, and disposal. For all numbers given from hereon see the table on page 111). So, today's 70-km-car ride will cause 9.1 kg of direct CO_2 emissions. Taking the train instead, I would emit only 0.35 kg. That is a ratio of 26:1! (For a comparison of direct GHG emissions of different means of transport, see page 111.) Now, since CO_2 emissions are not really an "externality," and I need to add their cost to the total for the ride, the question is how to price them. On the one hand, in *what form* to price them; on the other: *how high* to price them.

The first aspect is easy to tackle: You simply put an additional price on fossil fuels whose consumption causes CO_2 emissions. This additional price then of course applies to all kinds of fossil fuel consumption, not only for driving a car (but also for heating, food, and everything else causing emissions). In other words: you introduce a **carbon tax**. The basic concept behind such a tax is that it puts a price on any polluting activity causing the emission of CO_2. The individual, company, or organization causing that pollution—which all people and the environment will suffer from—pays a tax, or rather a fine for doing so. This is the so-called *polluter pay principle*. It should also—or even more so—apply to the companies taking fossil fuels out of the ground and onto the market. This extraction is the first kind of pollution happening.

Tackling the second aspect as to how high this tax should be is when it gets tricky. For one, how do you calculate the cost of damages caused by raised atmospheric CO_2 levels? Such calculations are being undertaken—by the insurance industry, for example—and albeit complex in nature (as they involve global phenomena), they usually quickly reach cumulated damages over decades in the hundreds of trillions of Euros! For our purposes here, let's stick with easy-to-grasp real-world examples from countries where a carbon tax already exists—in one third of European countries, as a matter of fact (even though the tax is marginal in most cases). I take Sweden as an example: There, carbon taxation was already introduced in 1991, and currently the price for one tonne of CO_2 emissions is 110 Euros.[50] There we have it then: Following the Swedish example, the carbon tax on my 70-km-car ride, resulting in 9.1 kg of CO_2 emissions, would add exactly one Euro to costs (and four Cents for the train ride).

Externalities

ARE COSTS INCURRED BY AN ECONOMIC ACTIVITY BUT PAID FOR BY NO ONE: LIKE ENVIRONMENTAL POLLUTION THROUGH EMISSIONS. FOR DECADES, SUCH COSTS WERE SIMPLY DUMPED ON NATURE.

TODAY'S TRIP

70 km back and forth

BY CAR

Cost: 24–41 Euros

**Emissions: 9.1 kg CO_2
(and other pollutants)**

Pollution tax: none

The carbon tax for 9.1 kg CO_2
emissions should be at least 1 Euro.

BY TRAIN

Cost: 9.60 Euros

Emissions: 0.35 kg CO_2

Pollution tax: none

The carbon tax for 0.35 kg CO_2
emissions should be at least 4 Cents.

Photo: iStock.com / Wenjie Dong

Waste in the mobility system*

50% of city land is dedicated to streets and roads, parking, service stations, driveways, signals, and traffic signs (on average).

5% of the time roads reach peak throughput (i.e., register the highest number of cars). But even then only 10% of roads are covered with cars.

92% of time the typical European car is parked.

A **12:1** average dead-weight ratio and

86% fuel inefficiency make fossil fuel-powered cars a highly wasteful form of mobility.

* Based on: ELLEN MacARTHUR FOUNDATION (2019).
Photo of Berlin in the background: iStock.com / Tomeyk | Photo of car: Christian Schienerl

I believe I could still easily afford the car ride even if it was properly car-bon-taxed. But let's look at the tax for a whole year: Doing 6,000 km with my car causes 780 kg of CO_2 emissions. The carbon tax for that would be 86 Euros—still not a lot, if you ask me. My car's relatively low emissions are due to its relative lightness (1,215 kg). While that vehicle weight already equals carrying around ca. 17 times my bodyweight as deadweight whereever I drive, the ratio gets even more absurd if I imagine myself in one of the now so popular heavy *sports utility vehicles* (SUVs). Weighing 2,000 kg (27 times my bodyweight) or more and emit-ting 260 g of CO_2 per kilometer (twice that of my car), total emissions for the same 6,000 km in a year would now be 1.56 tonnes. While the passenger has remained the same, this car now carries even more deadweight around, conse-quently causing more emissions. The carbon tax for driving 6,000 km in that SUV would be 172 Euros. Judging by Austrian income levels, that is still quite a small amount.

In political discussions the case *against* a carbon tax is often made by in-voking the example of low-income commuters who have to travel to work each day by car because they don't have proper public transportion where they live. A carbon tax would unfairly handicap such commuters it is maintained. Let's do the math, be generous, and say a commuter has to do a 200-km-round-trip to and from work each of the five workdays in a week. That commuting would result in 1,000 km driven per week. Multiplied by 47 workweeks a year, that makes 47,000 km per year for the work commute! If the commuter drives a small car like mine, this commuting is going to cause 6.1 tonnes of CO_2 emissions over the whole year. An Austrian citizen, on average, causes 7.2 tonnes per year altogether. I would call this highly disproportionate then: Does this commuter really have to cause almost as many CO_2 emissions per year as an Austrian does altogether, on average? But let's say that this is the case: So the commuter would have to pay 671 Euros per year (or 56 Euros a month) in carbon tax. As the argument goes, let's agree that this tax would disproportionately handicap the low-income commuter. You should remember that they also have to pay carbon-tax on other things like heating, food, and so on. But how is that a proper argument against a carbon-tax? Just as there are other tax breaks for low incomes, the same would apply to a carbon tax: the low-income commuter could simply be exempt from it. But for all others driving their light- or heavyweight petrol or Diesel vehicles to work or in their spare time, the polluter-pay-principle should apply: the more pollution, the more tax. No pollution, no tax—or maybe even a bonus for get-ting around eco-friendly? In a very simple carbon tax model that could be easily implemented: the state proceeds from polluting citizens' carbon tax could go to other citizens causing no pollution as an eco-bonus. In the grand scheme of things, this would create a *steering effect* away from polluting toward non-pollut-ing behavior. Me? I would be happy to pay carbon tax for the 780 kg of emissions per year I cause by driving my car. As there currently is no carbon taxation in Austria, I do so voluntarily, by offsetting these emissions via goldstandard.org.[51]

But as I am trying to stop global heating by tonight there is a more important question to answer right now: Instead of just offsetting emissions through a tax I pay, **how can I** *avoid* **as many of the emissions related to my mobility as possible?** Here are some of my thoughts and temporary solutions for this: 1. Most of the time in the city I leave my car at home anyhow, drive my bike, walk, or use public transportation. 2. I could get rid of my car altogether, rent/share a vehicle, or take a cab when in need of individual automotive transportation (options available mostly in cities). I have done the math: this would cost a fraction of what it costs to keep my own car. 3. Since I like driving and being behind the steering wheel of my own car (don't ask why, because I don't know): how about doing it almost emissions-free? Yes, I could switch to an **electric vehicle (EV)**. I very much like the idea of driving around emissions-free locally (meaning wherever I go). In Austria, with its high ratio of renewably generated electricity operation of an EV would make a lot of sense. Much less so in countries like Poland where a huge portion of electricity is still generated by fossil-fuel-powered plants. Apart from the *use* of an EV—which is more or less emissions-free (depending on how the electricity going into the battery is generated)—there is a huge downside to this type of vehicle though: Production of EVs—especially of their lithium-ion batteries—has a much higher Ecological Footprint than that of PEVs. For a life cycle analysis of different types of vehicles, covering their production and operation, see page 111 →. Buying a new EV now would set me back ca. nine tonnes of CO_2 emissions embodied in its production. For that amount of emissions I could go on driving my petrol car for almost another twelve years. In terms of my Ecological Footprint NOW, maybe it would be best then to just keep my relatively economic petrol car and drive it even less?

Momentarily, I arrive at the printers. With a nod in agreement, I notice the EV charging station, which the printers have installed in their parking lot. Maybe a few years from now, I will charge my recycled battery electric vehicle (its production having caused far less emissions than today) at that station while being here for hardproofs of books, its electricity coming from the massive windpark I just drove through. Or maybe—to keep things really simple—I will just take the commuter train here more often in the future . . .

DIRECT GREENHOUSE GAS EMISSIONS OF DIFFERENT MEANS OF TRANSPORT

(during operation)

Average greenhouse gas emissions in grams CO_2 equivalent per *passenger kilometer* (moving one passenger a distance of one kilometer) or *tonne kilometer* (moving one tonne of freight a distance of one kilometer)

ROAD

Petrol engine car
146.4   (grams CO_2e/km)

Diesel engine car
148.6

Battery electric vehicle
0.0

Light-duty Diesel van (deadweight < 3.5 t)
641.7

Heavy-duty Diesel truck (< 18 t)
290.7

Heavy-duty Diesel truck (> 18 t)
112.1

Diesel tour bus
36.5

Public bus (drivetrain mix)
41.3

RAIL

Passenger train
5.0

Freight train
2.4

AIR

Airplane, short-haul (< 3 hrs flight time)
820.1

Airplane, medium- & long-haul (> 3 hrs)
386.0

Graphic: SCHIENERL D/AD; Data source: Österreichisches Umweltbundesamt (Environment Agency Austria), retrieved Aug. 7, 2020, from: https://www.umweltbundesamt.at/fileadmin/site/themen/mobilitaet/daten/ekz_pkm_tkm_verkehrsmittel.pdf

LIFE CYCLE ANALYSIS OF DIFFERENT CAR TYPES

(total GHG emissions for production, operation, and end of life)

The table below compares total *direct* GHG emissions from *operation and recycling* and *indirect* emissions from *production*. In Austria, with its high ratio of renewable electricity generation, emissions of a small electric vehicle are low (see "Battery electric vehicle" with Austrian power mix). The only two shorter bars show technologies not yet commercially developed: hydrogen (H_2) from electrolysis and synthetic Diesel, both generated by wind power.

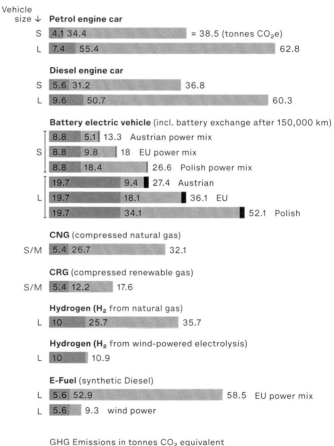

Vehicle size ↓

Petrol engine car
S   4.1 34.4   = 38.5 (tonnes CO_2e)
L   7.4 55.4   62.8

Diesel engine car
S   5.6 31.2   36.8
L   9.6 50.7   60.3

Battery electric vehicle (incl. battery exchange after 150,000 km)
S   8.8 5.1 13.3 Austrian power mix
S   8.8 9.8 18 EU power mix
S   8.8 18.4 26.6 Polish power mix
L   19.7 9.4 27.4 Austrian
L   19.7 18.1 36.1 EU
L   19.7 34.1 52.1 Polish

CNG (compressed natural gas)
S/M   5.4 26.7   32.1

CRG (compressed renewable gas)
S/M   5.4 12.2   17.6

Hydrogen (H_2 from natural gas)
L   10 25.7   35.7

Hydrogen (H_2 from wind-powered electrolysis)
L   10 10.9

E-Fuel (synthetic Diesel)
L   5.6 52.9   58.5 EU power mix
L   5.6 9.3 wind power

GHG Emissions in tonnes CO_2 equivalent
Vehicle size: S = small, M = medium, L = large

Production | Operation and recycling
| Duration of operation: 15 years (13,000 km/year)

Reduction by 0.6 tonnes CO_2e through reuse of materials | Reduction by 1.7 tonnes CO_2e through reuse of materials

Graphic: SCHIENERL D/AD; Data source: ÖAMTC (Austrian Automobile, Motorcycle and Touring Club), *auto touring*, Sept. 2019, p. 19

Lithium harvesting in the salt flats of Salar de Uyuni, Bolivia, 2019. This region bordering on Bolivia, Chile, and Argentina is said to hold 70% of worldwide lithium deposits. Flushing out the lithium alloy from the salt body requires up to 80,000 litres of freshwater per hour, a resource particularly scarce in this region. Groundwater gets increasingly depleted and contaminated, leaving indigenous subsistence farmers in the region high and dry and faced with new toxic compounds killing their livestock. Production of lithium battery packs is also highly energy-intensive, which explains the high production footprints of EVs (see table on p. 111). Image: NASA Visible Earth

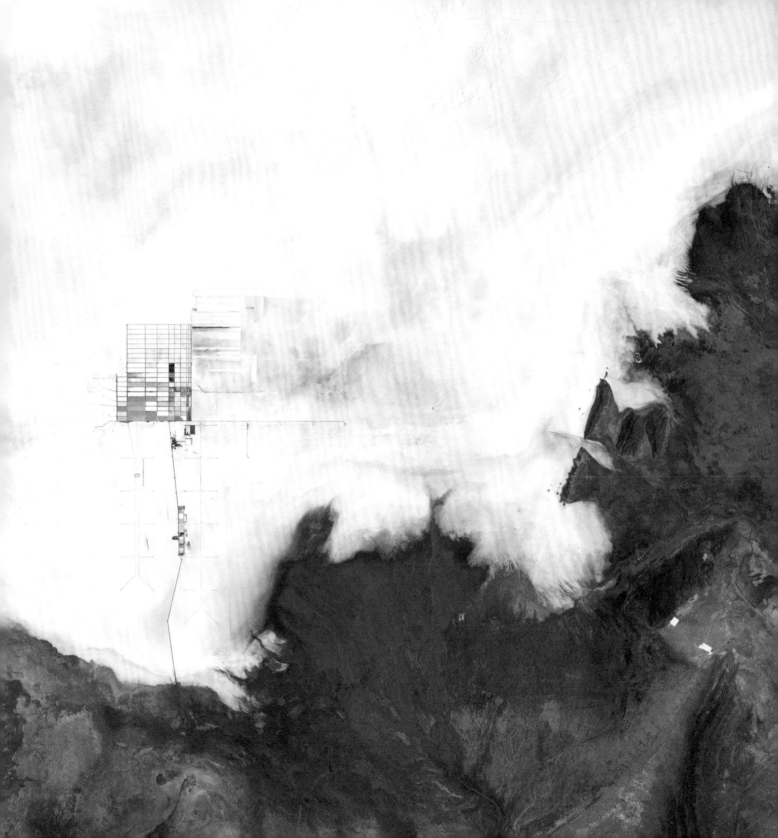

Production hall of a printing company.
A high-precision sheet-fed printing press
with eight (+1) color-inking stations is
visible. This is the press, which this book
was printed on. Photo: gugler/RitaNewman

AT THE PRINTERS
So, *that's* industry!

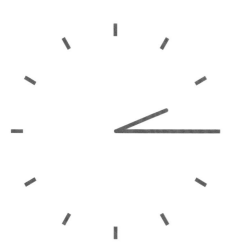

The printer greets me as I arrive, just in time to check the first print sheet. I have known him for some time now, and we have printed quite a number of books together, most of them on this printing press, which they have had for a few years now. It is a state-of-the-art, high precision sheet-fed machine measuring 20 meters in length. Everything about this rig—from the sheet-feeder at one end and the eight color-inking stations to the printing desk at which we are standing—is geared for precision and speed. It can print 16,000 double-sided sheets per hour, each measured for minimal spatial and color aberrations inline, which get corrected automatically during printing. The whole assembly weighs 30 tonnes, consists of thousands of moving parts, and is controlled by software that monitors the movement of the print sheet at a spatial scale of tenths of millimeters.

The result is just text, photos, and graphics on paper: quite a mundane product. But it's done brilliantly by this machine, which is the best in its class—or put differently: it's the industry standard. It embodies perfectly what **industry** is all about: standardization, precision, efficiency, and speed. Industry's ultimate purpose is to produce large volumes of highly similar items in the shortest possible time, with possible error sources reduced to a minimum.

During early industrialization this meant getting rid of slow and error-prone human handwork. Humans are not good at performing repetitive tasks with precision at high speeds. That is what weavers in late-eighteenth-century England understood right away, when new mechanized power looms, driven by coal-fired steam engines, were introduced and rapidly ended weaving and cloth-production as a "cottage industry." This first industrial machinery, powered by James Watt's patented refinement of the steam engine from 1769, initiated the first wave of industrialization in Great Britain known as the **Industrial Revolution**.

Steam-powered machines and mechanization introduced a new rhythm and intensification of time in human economy. Human and animal labor became a subordinate function, merely supporting the industrial machine's operation. No wonder a movement of angry workers, called **Luddites**, quickly formed and took sledgehammers to this new machinery in acts of rage. The Luddites were not technophobic per se, they just took out their anger about becoming obsolete on the very things that had replaced them. Those who were going to operate and support the new industrial machinery would soon form a new class of laborers: Recruited from still largely rural populations, driven into towns and cities, where the newly established factories were situated (leading to a massive wave of urbanization), these laborers—men, women, and children—would become the wage-dependent **industrial proletariat**. Early socialist thinkers like Charles Fourier and, later, Marx and Engels expounded on how a new class of **capitalist industrialists**—the ones owning the means of production and factories—eagerly embraced the new and more efficient way of turning human labor into *added value* embodied in products and goods.

The new industrial mode of production was first employed in Great Britain's textile and steel sectors but rapidly spread from there to other lines of business, to continental Europe—first to Germany, Belgium and France—and then also to the US, Japan, and Russia. At the same time, the socialist movement grew, its rallying cry "Workers of the world unite!" (from Marx and Engels's 1848 *Communist Manifesto)* echoing from all newly industrialized corners of the world.

Not only did the new mode of production spread rapidly, it would also drive enormous economic growth and expansion in the decades to come. **Between 1820 and 2015 the world economy grew 90-fold** (!). During the first wave of industrialization, national economies' demand for raw materials was still met by imperial colonialism, the first manifestation of globalization, which had developed since the early sixteenth century.

Yet the new engines of industry also required much more energy than had been harvested from water, wind, and animal or human muscular power (think of slavery) during previous centuries. "The Anthropocene began around 1800 with the onset of industrialization, the central feature of which was the enormous expansion in the use of fossil fuels."[52] **Cheap, high-energy fuels were the resource most crucial to the success of the industrial process:** Coal, oil, and later gas fueled the transition to a high-energy society. **Worldwide fossil fuel consumption grew a mind-boggling 850-fold between 1820 and 2015** (!) (see p. 83).

While the first Industrial Revolution was powered exclusively by the coal-fired steam engine, by the end of the 1860s another fossil fuel took center stage. First in Poland, and shortly thereafter in Pennsylvania in the US, **the first oil wells** were dug. What followed was one oil boom after the other. In the US, these booms were closely linked to the expansion of the coal-fired steam railroad from the East to the West coast. But fossil-fuel-driven mobility was soon to become independent of rails even. The second wave of industrialization was already well underway, when the invention of the **internal combustion engine** (the "Otto engine," patented and first produced in 1877) drove developments, which led to the first industrially produced cars (e.g. the Ford T "Tin Lizzie," of 1908). Around the same time, in 1896, Swedish physicist, chemist, and later Nobel Prize laureate Svante Arrhenius first published the theory of a "greenhouse gas effect."[53] In later works, Arrhenius also pointed out that this natural greenhouse effect would be boosted by the increasing combustion of fossil fuels by humans.

While my thoughts have drifted off to the beginnings of the industrial era and the importance of fossil fuels in the expansion of an industrialized world economy, the contemporary industrial machine I am standing at has already printed 10,000 sheets, each copy indistinguishable from the others. While this is true for the printing, it is also true for the high-quality paper we use for this book. It is produced in Sweden, at the paper mill in Munkedal, about an hour north of Gothenburg. I once visited the mill and left it quite impressed with the even more

THE NEW
engines of industry
REQUIRED MORE ENERGY THAN HAD PREVIOUSLY BEEN HARVESTED FROM WATER, WIND, AND MUSCULAR POWER . . .

Rotative steam engine by Boulton & Watt, 1788. Built by James Watt in 1788, this coal-fired steam engine was used at Matthew Boulton's Soho Manufactory in Birmingham, where it drove 43 metal polishing machines for 70 years.

Photo: Science Museum Group Collection Online. https://collection.sciencemuseumgroup.org.uk/objects/co50948.

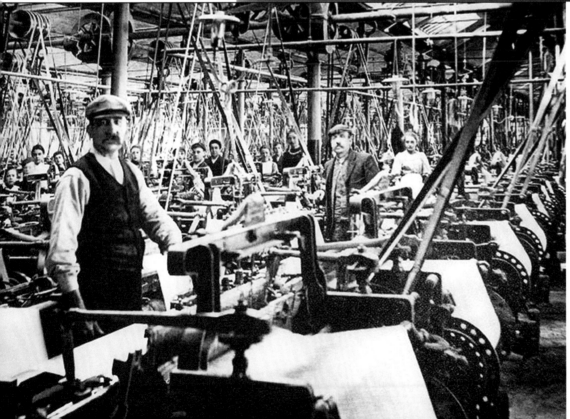

Workers at a textile mill, last quarter of 19th century, probably Great Britain. Men, women, and children standing at the steam-powered looms they operate. The looms are driven by transmission belts suspended from the factory ceiling. A centralized, coal-fired steam engine drives the belt-system.

Photo: source unknown

Smokestacks of the Krupp steelworks rising from the city of Essen, Germany, 1890.

Photo: source unknown

Assembly line for the Ford Model T ("Tin Lizzie"), the first mass-produced petrol engine vehicle, 1920s.

Photo: Alamy Stock Photo

COAL, OIL, AND GAS
STEPPED IN AND DROVE
ENORMOUS ECONOMIC
EXPANSION DURING THE
NEXT 200 YEARS.

gigantic industrial milling machinery they have in use there. Not only is the paper it produces high-quality, it is also of consistent quality. This means that whichever batch we print on today is going to be indistinguishable from the same brand paper I used a few weeks or a few months ago. Again: this product bears all the hallmarks of an industrial mode of production.

But come to think of it: This paper has travelled quite a long way to get here from Sweden. The same holds true for the printing inks and aluminum printing plates, which are manufactured in Germany. Basically all of the materials and intermediate products we use here to print and bind a book as a final consumer product were shipped here from someplace else. That, it seems, is another characteristic of today's manufacturing industries: The materials they use come from all over the world, get assembled in one place (or multiple locations), then get shipped to their final points of consumption, which are again spread out over the whole globe.

Just as I think of this, I receive a call on my mobile phone. Though its display is partly broken, I can see that Nina is calling. She notifies me that she just accepted a package in my name at the office. I thank Nina and tell her to take the rest of the day off if she has run out of stuff to do. While I check the next print sheet, I think of that package at the office. I know exactly what's inside: It is my new smartphone, which is going to replace the broken and outdated model in my backpocket whose battery only lasts a couple of hours anymore before it needs recharging.

Industry 4.0
Late-twentieth-/early-twenty-first-century industrial practice increasingly employs automatization and digitization. Human labor is supplanted whereever possible. The image shows a car manufacture assembly line at a Chinese factory dominated by industrial robots.

Photo: iStock.com / xieyuliang

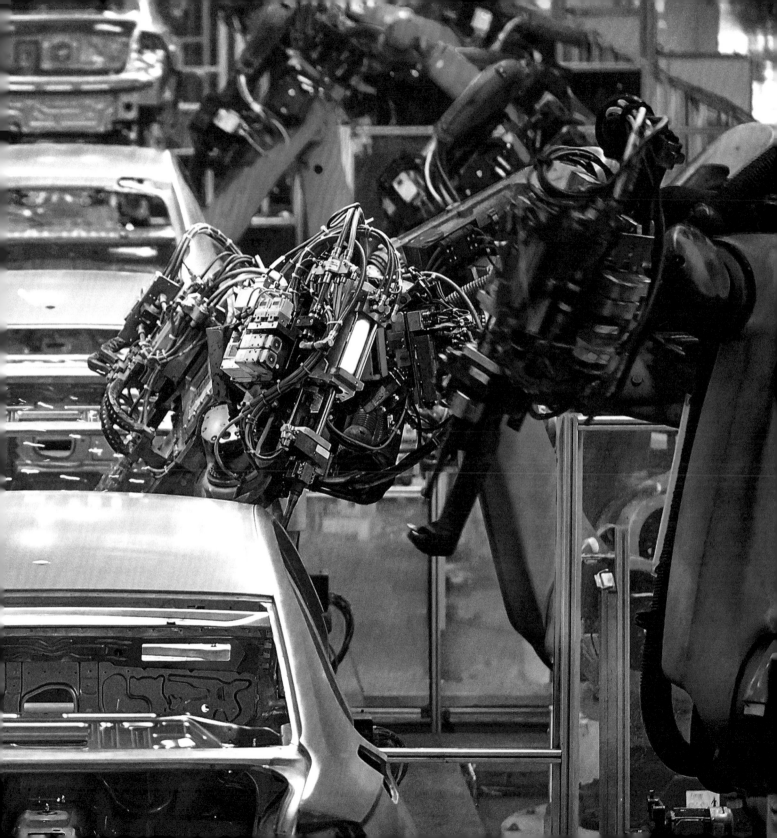

The contents of this package had to circumnavigate Earth multiple times before they could be delivered to my doorstep so conveniently.
Photo: iStock.com / katleho Seisa

DELIVERY OF MY NEW SMARTPHONE

A globalized and greatly accelerated economy

While we go on printing the book, the package that Nina received from the courier now quietly sits on my desk at the office. It may be a good thing it can rest there for a while after all its travels: Before delivery to my doorstep half an hour ago, the courier had taken it on a mini-truck-ride through Vienna among many other packages he has to deliver today. Last night, the package had arrived at a logistics center close to the city, where it was sorted and then put on that courier's delivery list for today. Yesterday, the package had travelled hundreds of kilometers on a truck from Germany en route to Vienna. The night before last, it had been delivered to a German logistics hub by truck, having originated at a storage facility in the Netherlands. Twelve hours before that, conveniently sunk into my sofa with my laptop on my lap, I had ordered the smartphone on an on-line trading platform. For a few days, the smartphone had been sitting at the storage facility in the Netherlands among thousands like it. Prior to that, whole containers full of boxes containing smaller boxes with smartphones like mine in them had been cleared by customs at Europe's largest port in Rotterdam. Those containers had arrived by freight ship from China a day earlier.

I could go on reverse-engineering what travels my new smartphone has been on before it was assembled half-way around the globe in Shenzhen, China. We could backtrack where intermediate products going into that final assembly had come from. We could probably determine where on the globe materials going into those intermediate products had been mined (see p. 134). But for now, let it suffice to say that we all are very familiar with phenomena like the ones described above under a common denominating term: globalization. We now all live in "One Big World on one small planet,"[54] the parts of which are constantly connected through international trade, freight traffic, travel, telecommunications infrastructures (not least for the Internet), global financial markets, military and civil infrastructure like GPS, transoceanic data cables, and so on.

The idea for a truly global marketplace is much older and more pervasive though than the digitized iteration we are all so familiar with today, upheld by Amazons and Alibabas. The notion of a globalized world economy originated in the mid-1940s, under the impression of the three major international crises, which had dominated the preceding decades: the First World War (WWI), the Great Depression, and the Second World War (WWII). Before WWI, a first wave of globalization with increasing volumes of international trade had already occurred from ca. 1860 onwards. But that first wave came to an abrupt halt when European powers entered into a four-year-long war with each other (1914–1918), joined by the US in 1917. After that war, the political face of Europe had entirely changed: Previous superpowers like the Austro-Hungarian Empire were dissected into multiple nation states, monarchies ended—most notably in Russia after the 1917 October Revolution. In the aftermath of WWI, with tense economic conditions mounting all over Europe, fascism rose as a new political power in the nations that had lost WWI. Meanwhile, across the Atlantic, the US slipped into the worst economic crisis of modern times and a multiple-year-long recession that had been brought about by unharnessed financial markets and the stock market crash in 1929. The environmental crisis of the Dust Bowl added further strain on the American economy. Fascist movements took power in Italy, Germany, Spain, and Austria and led the whole continent into another great war and one of humanity's darkest historical times. A brutal six-year-long total war was overshadowed by even more gruesome industrially organized genocide—the Holocaust. At least 60 million people died in WWII, at least six million of those European Jews, who were killed in and outside of Nazi death and concentration camps.

Deeply shocked by the cataclysms which imperialist politics, fascism, and unchecked capitalism had brought over the world, after WWII a newly forming "world community" pondered ideas of how to avert such catastrophes in the future. One consequence of such deliberations was the establishment of the United Nations in 1945 and the proclamation of the Universal Declaration of Human Rights (UDHR) in 1948. One other prominent idea of the time how to prevent desastrous wars in the future was the (neo)liberal theory of free markets—an "invisible hand" regulating not only prices of goods, services, and wages but also guaranteeing a stable political constitution of the societies built upon such market economics (that liberal idea had first been expressed by economist Adam Smith at the end of the eighteenth century). The Mont Pèlerin Society, founded in 1947 by Austrian economist Friedrich August von Hayek, quickly became the central intellectual hub of a newly forming neoliberal network in Western Europe and the US. In opposition to a recently formed bloque of communist countries under the wings of the Soviet Union, in the West another model was pursued: liberal democracies resting on the foundations of capitalist market economies. The more nations would be engaged in interactions with each other via markets and international trade, the less likely they would be to ever wage war against each other again—such was the reasoning of neoliberalist thinking. This promotion of international peace was to be a mere side-effect of what neoliberalism propagated though.

As a consequence of the Bretton Woods Conference in 1944, which had established a system of fixed currency exchange rates between signatory nations, the General Agreement on Tariffs and Trade (GATT) was put into effect at the beginning of 1948. As a legally binding international treaty it was meant to solve market-related disputes between signatory nations.

From the 1950s onward, global economic growth (as expressed in real *gross domestic products* or GDPs) and the expansion of international trade (as expressed in the volume of exported goods and services [see graph ↗]) were intricately linked and on a common exponential growth path. The same was true for the emerging industrial agriculture and a soaring world population. All these socioeconomic developments are illustrated by the graphs on page 130: Most of

WE NOW LIVE IN
One Big World
on one small planet,
CONSTANTLY CONNECTED
BY TRADE, FREIGHT,
TRAVEL, FINANCE,
ANALOGUE AND DIGITAL
INFRASTRUCTURES.

VALUE OF GLOBAL EXPORTS

"The integration of national economies into a global economic system has been one of the most important developments of the last century. This process of integration, often called Globalization, has materialized in a remarkable growth in trade between countries.

The chart here shows the value of world exports over the period 1800–2014. These estimates are in constant prices (i.e. have been adjusted to account for inflation) and are indexed at 1913 values [1913 = 100%] . . . Exports today are more than 40 times larger than in 1913."

Graphic and text: OurWorldInData.org/trade-and-globalization / Source: Fecerico and Tena-Junguito (2016)

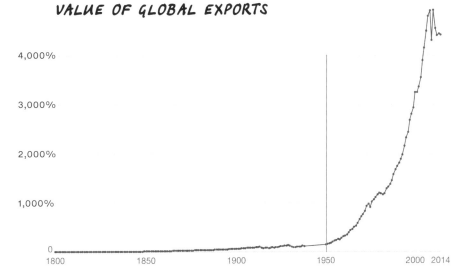

Hong Kong
freight terminal
Photo: iStock.com / yupiyan

The globalized world
Earth is covered by a densely woven net
of cities (yellow), roads (green), shipping
lanes (blue), flight routes (red), railways,
transmission lines, pipelines, submarine
cables, and other global infrastructure.
Image: NOAA / Science Photo Library / APA Picturedesk

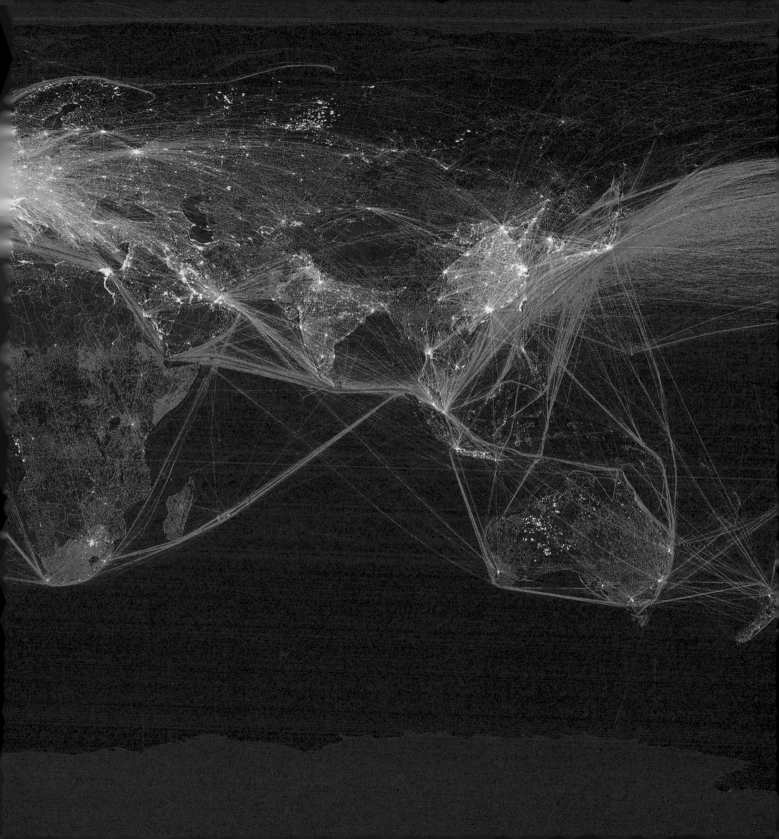

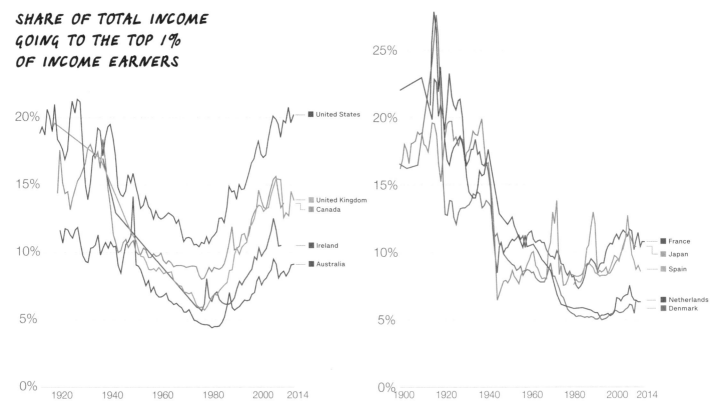

SHARE OF TOTAL INCOME GOING TO THE TOP 1% OF INCOME EARNERS

Left chart (English speaking countries):

20%
15%
10%
5%
0%

1920 1940 1960 1980 2000 2014

- United States
- United Kingdom
- Canada
- Ireland
- Australia

In English speaking countries inequality followed a U-shape

Right chart (continental Europe and Japan):

25%
20%
15%
10%
5%
0%

1900 1920 1940 1960 1980 2000 2014

- France
- Japan
- Spain
- Netherlands
- Denmark

. . . whereas in continental Europe and Japan it followed an L-shape.

← **UK Prime Minister Margaret Thatcher** (1979–1990) and **US President Ronald Reagan** (1981–1989) at the House of Commons, London, in 1978. *Thatcherism* in the UK and *Reaganomics* in the US ushered in the ongoing era of turbo-charged neoliberal free-market policies in Western democracies. Photo: KEYSTONE Pictures USA / Alamy Stock Photo

↖↑ What that meant for **income inequality** can be seen in the graphs above. In English speaking countries (left) inequality has soared again since the beginning of the 1980s: In the US today, 20% of income go to the top 1% of income earners.

Graphic: OurWorldinData.org/income-inequality. Source / author: Max Roser (CC-BY-SA)

Neoliberalism

IS FOUNDED ON THREE PILLARS:

1. PRIVATIZATION & PUBLIC SPENDING CUTS

2. DEREGULATION OF THE CORPORATE SECTOR

3. TAX CUTS FOR CORPORATIONS AND HIGH INCOMES

these graphs show slow but pronounced linear growth until WWII, which then accelerates into exponential growth from the 1950s onward. This latter phase (1950s to the present) has therefore become known as the Great Acceleration—the second phase of the Anthropocene (whether it is still lasting or has already come to an end, is a matter of ongoing debate).[55]

The graphs on page 131 illustrate how these socioeconomic developments are mirrored by changes to the Earth System: What is shown in these graphs are exclusively processes of over-exploitation, degradation, and depletion.

These two sets of statistics clearly show that the Great Acceleration and expansion of globalized economies, accompanied by gains in material welfare for ever larger portions of a burgeoning world population, were founded for the most part on the exploitation and degradation of natural resources and ecosystems—not least through the destructive and unsustainable overuse of energy derived from fossil fuels. The post-WWII decades have often been framed as the *Economic Miracle* years. Economies in Western industrialized countries were booming, middle classes in these countries were expanding, everything seemed possible—even putting men on the moon (see intro). But all the while, the world was incurring mounting debt: not only of the monetary kind but especially in the living environment, within which its economies are operating. First and foremost, the fossil fuel energy powering that economic boom had serious side-effects right from the start, which came into focus only much later.

Fittingly, the first big shock to the newly globalized economy came with the first oil crisis in October 1973 when OPEC (then: "Organization of Arab Petroleum Exporting Countries") imposed an oil embargo on nations it believed to be supporting Israel in the Yom Kippur War (the US, the UK, Canada, the Netherlands, and others). Another oil shock followed in 1979. After both oil crises international recessions ensued.

As a remedy to these recessions a new wave of radicalized neoliberalist ideology took hold of market economies during the early 1980s. *Thatcherism* in the UK and *Reaganomics* in the US paved the way for a turbocharged neoliberal regime being founded on three pillars: 1. *privatization* and erosion of the public sector (cuts in welfare state public spending), 2. *deregulation* of the corporate sector, and 3. corporate and high-income *tax reductions*.[56] It was this third wave of (neo)liberal market economics that led to the dismantlement of large portions of the manufacturing industries in the West and their reestablishment in low-wage, low-regulation Asian countries like (first) Korea and (then) China. It also brought about the degradation of Western public sectors, including infrastructure and social security spending cuts. Furthermore, virtually unchecked financial markets were allowed to emerge, and they quickly exceeded outputs of real economies by a multiple. The establishment of corporate tax havens all around the world allowed transnational corporations to avoid taxation in their

THE ANTHROPOCENE: SOCIOECONOMIC TRENDS

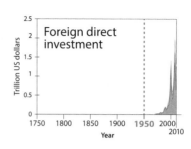

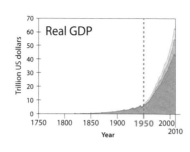

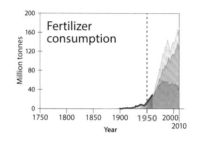

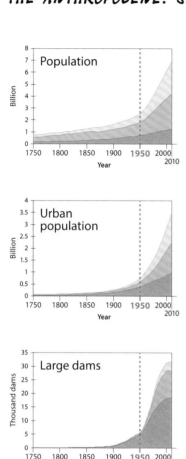

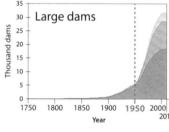

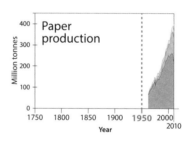

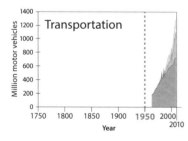

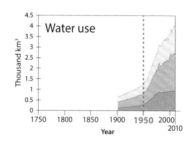

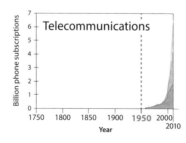

↖ The set of graphs on this page shows several **socio-economic indicators of human development** during the last 270 years. The data is differentiated by high-income **OECD countries** (EU-28, Australia, Canada, Iceland, New Zealand, Switzerland, Turkey, USA), the emerging economies of the **BRICS countries** (Brazil, Russia, India, China, South Africa), and **other countries**.

↗ The set of graphs on the right page shows **Earth System trends** during the same period. These are almost exclusively developments of overexploitation and degradation of nature (atmospheric concentrations of **carbon dioxide** are given in ppm [parts per million], those of **nitrous oxide** and **methane** in ppb [parts per billion]; **ocean acidification** is measured in nanomoles per kg of hydrogen ions, **nitrogen to coastal zone** in million tonnes per year of human-caused nitrogen flux).

OECD ▨ BRICS ▨ Others

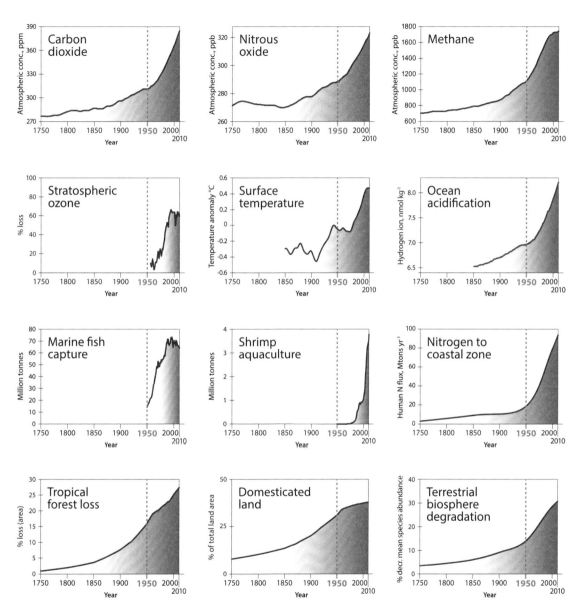

own countries. The constitution of the World Trade Organization (WTO) in 1995, which succeeded GATT and enforced a powerful and effective global trade regime, finalized the neoliberal reorganization of the globe.

Also recently established (1988), the United Nations Framework Convention on Climate Change (UNFCCC) and its Intergovernmental Panel on Climate Change (IPCC) were pitted against the interests of this new global trade regime right from the start, as Naomi Klein has powerfully demonstrated.[57]

During this most harmful phase of the Anthropocene between 1970 and 2020, mankind finally overwhelmed the great forces of nature and became the single most powerful agent of change in the physical world. Not least in that sense, the human and the natural worlds are now entirely merged and globalized: whether some of us like it or not, we now inhabit "One Big World on one small planet as One People of Earth citizens."

My new smartphone is the perfect illustration (see pp. 134–135): designed in California by the world's second most valuable company, made up of multiple functional parts patented under US laws, assembled in China under more or less regulated working and environmental conditions from parts manufactured in diverse, mostly low-income countries under more or less regulated working and environmental conditions with raw materials, some of them *conflict minerals* and *rare earths*, mined in even more low-income countries of the world under even less regulated working and environmental conditions, transported between those geolocations by a constant stream of freight traffic by air, sea, and land, and ultimately delivered to me here in high-income Austria by a globally operating online trading platform. This device truly is a global venture and product. If you were to trace all the economic, social, and environmental impacts of its production, consumption, and recycling (or disposal as tech waste) you would end up with a complete picture of that global human enterprise and its darker sides that have brought forth such a product. The laconic information given by the manufacturer/brand is that the device's carbon footprint is 72 kg of CO_2 emissions for all stages of its life cycle (production, shipping, use, disposal/recycling).

While I have been lost in reveries about the origins of my new smartphone, we have finished printing the book. In a rush I bid goodbye to the printer and exit the production hall, being swallowed by the cold winter air of this late January afternoon.

Bingham Canyon mine, Utah, USA, 2018.
At 4 km width and a depth of 1.2 km this
is the world's largest open-pit mine and
the deepest man-made excavation on
Earth. Since mining began in 1906, more
than 19 million tonnes of copper have
been extracted from this site.
Photo: Eric Prado (CC BY-SA 4.0)

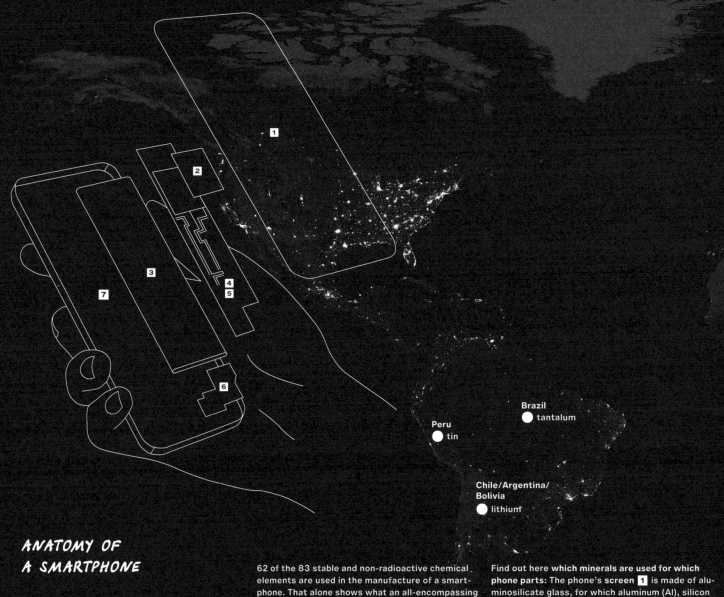

Brazil
● tantalum

Peru
● tin

Chile/Argentina/
Bolivia
● lithium

ANATOMY OF A SMARTPHONE

62 of the 83 stable and non-radioactive chemical elements are used in the manufacture of a smartphone. That alone shows what an all-encompassing venture today's electronics are. But their production also encompasses almost all of the globe, as well as the social, political, and ecological problems of countries, where mining and production takes place: See the map to find out where ● elements (or the minerals containing them) are mostly mined, and where different phone parts are ■ manufactured, tested, and assembled before devices get shipped all over the world to consumers.

Find out here **which minerals are used for which phone parts:** The phone's **screen 1** is made of aluminosilicate glass, for which aluminum (Al), silicon (Si), and potassium (K) are needed. Its touch capabilities are due to a thin film of transparent circuits made of indium (In), oxygen (O), and tin (Sn), which is a *conflict mineral*. The mining of **conflict minerals** tin (Sn), tungsten (W), gold (Au), and tantalum (Ta) in war-ridden countries and conflict zones (e.g. in the eastern Democratic Republic of Congo [DRC]) often directly finances the perpetuation of fighting. The screen's brilliant colors are made possible by *rare earths* yttrium (Y), europium (Eu), terbium (Tb), and gadolinium (Gd). **Rare-earth elements** (also: *rare-earth metals*) mostly come from China

Russia
● platinum-group metals, potassium

Belarus
● potassium

China
● germanium, graphite, indium, rare earths, sand, silicon, silver, tin, tungsten
■ SDRAM, final assembly

South Korea
● indium
■ SDRAM

India
● graphite, tantalum

Taiwan
■ chip fabrication, SDRAM, final assembly

Myanmar
● tin

Japan/ Singapore/ Vietnam
■ SDRAM

Rwanda
● tantalum

Phillipines
■ SDRAM, packaging, final testing

DR Congo
● cobalt, tantalum

Indonesia
● tin

Australia
● lithium, rare earths

(80–85%) and Australia (15%). They share superior magnetic and conductive properties. But in mining, they are regularly found alongside radioactive elements, which is why their mining operations (would) require radioactive waste management.
The **processor chip** 2 is made mainly of silicon (Si) but also uses phosphorus (P), antimony (Sb), arsenic (As), boron (B), indium (In), and gallium (Ga).
The **lithium-ion battery** 3 is made of mostly cobalt (Co), lithium (Li), graphite (=carbon [C]), and aluminum (Al). **Cobalt** is mostly mined in the DRC. Child labor and hazardous mining conditions are regularly reported. Cobalt mining is tracked by some manufacturers of smartphones, and some require their subcontractors to participate in third-party audits.

But supply-chain transparency remains sketchy. **Lithium** extraction is fraught with its own set of socioecological issues (see p. 112).
Electrical connections 4 in the phone are made of gold (Au), silver (Ag), copper (Cu), tungsten (W), and tantalum (Ta). Three of these elements (Au, W, Ta) are *conflict minerals* (see above), as is the tin used besides copper and silver for the phone's **soldering** 5.
Speakers and vibration 6 Besides boron (B), nickel (Ni), and iron (Fe), *rare-earth elements* (see above) praseodymium (Pr) and neodymium (Nd) are used for high-powered magnets in the speakers and the vibration module of the phone.
The **case** 7 is made of aerospace-grade aluminum.

The last time we met (p. 102), we looked at my energy use at home and my food consumption. This time I will account for how much my mobility (like driving my car or flying to places) and my consumption of goods (like my new smartphone, clothes, or getting a haircut) contribute to my Ecological Footprint. We use the term "goods" broadly here: It includes both manufactured products as well as commercial services offered by businesses of all kinds.

As with the other two portions I have already accounted for, I touch down relatively far below the Austrian average [A] and slightly above the World average [B] for my mobility. In relative terms, it gobbles up a 17%-portion of my overall Footprint. I own a small combustion engine car, which I do ca. 3,000 km on for private puposes each year. That is about a quarter of what the average annual road performance of a car is gauged at. So while I do *own* a car, you might say I don't *use* it much. In addition to driving my car, I do circa 1,500 km per train each year, plus an average of 10 km a week on inner-city and regional public transportation. Within the city, I mostly ride my bike to get to places. If I have enough time, I walk. I try to avoid flying altogether these days, but at least one ten-hour roundtrip flight per year somehow remains unavoidable (see p. 157). I do carbon-offset this flight (we will hear more about offsetting a little further on) but still: That one flight per year sets me back about one tonne of CO_2 emissions—far more than driving my car all year for private purposes, which causes ca. 390 kg of CO_2. So while it may be true that aviation is responsible for only circa three percent of global greenhouse gas emissions, that proportion is much higher for me as an individual, upwards of 15 percent of my overall emissions in one year—for just one flight!

In terms of consumption of goods I am a mixed bag of nuts, you might say. Mostly, I am rather modest in my needs but there are some things I own excessive amounts of. Clothes do not belong in the latter category though. I buy new clothes when I have to replace something old and worn. Very rarely I buy them just because I feel the need to seem fashionable. Even then, I do buy high quality at high prices and trust that these pieces will last me a long time (which they regularly do, but not always). The same holds true for pieces of furniture, which I buy new only when I am in need of them. Most of the furniture I own I have had for years (if not decades), and a large portion of those pieces are vintage—you might even call them "antiques." I am sure they have had a number of previous owners before me. Consumer electronics are not an addiction for me either. I have

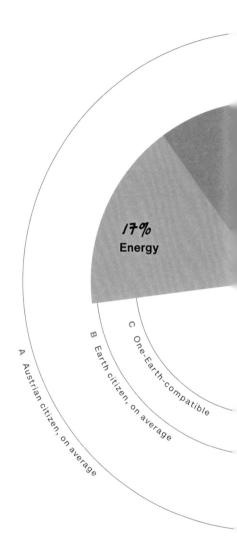

17%
Energy

A Austrian citizen, on average

B Earth citizen, on average

C One-Earth-compatible

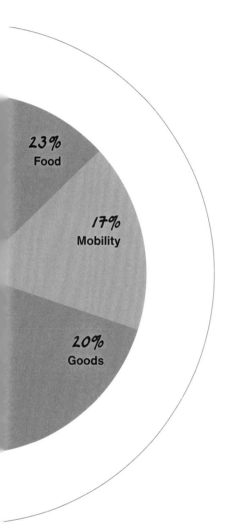

23%
Food

17%
Mobility

20%
Goods

had the smartphone that I now have to replace for more than six years. That's far longer than manufacturers would like to see me keep such a thing. Product cycles for consumer electronics have grown ever shorter over the past twenty years, and a lot of products are made to wear out quickly on purpose by manufacturers (this is so-called *planned obscolescence).*

I do own three computers, one tablet, and all kinds of digital camera and imaging equipment though, which I and my assistant Nina use for the services I provide as a freelance graphic designer. As I mentioned in the last footprint section, **work equipment** does not enter accounting of my *personal* Footprint. As any other resource that goes into my work, such *means of production* are embodied in the goods and services I offer, which others consume (just as the haircut I get shows up in my Footprint, and not in the barber's). The same is true for the used car, which I bought as my company car. Emissions and resources embodied in the car's production have entered my company's Footprint. As do the emissions and other costs from its operation for business purposes. But when I use the car for private purposes, emissions and costs for those kilometers driven enter my personal Footprint (see "mobility").

Something I have been doing in recent years regarding my company's Footprint is to **offset unavoidable carbon emissions**. Now, offsetting of course does not avoid emissions but is meant to compensate them by cancelling out emissions elsewhere in the world. You pay a certain amount of money per tonne of unavoidable emissions to an offset program provider (like goldstandard.org), and that money is then invested in projects (mostly in emerging economies) which build renewable energy capacity, capture methane from landfills or livestock, or distribute clean cooking stoves to families in poor regions. Offsetting of corporate or private emissions often sounds like *greenwashing* (trying to give the impression of acting environmentally friendly by doing something rather inconsequential), and often it actually is. But if it is done as a last step after having reduced one's own emissions to the level where they become unavoidable, then offsetting may be considered a good thing.

So, mobility and consumption of goods leave my ranking unchanged: I am still stuck halfway between an average Austrian Footprint and a sustainable one. I still fail to keep my promise to stop global heating. I am coming to suspect this failure may be "systemic". . .? Perhaps it also has something to do with the large quarter-portion of my Footprint still unfilled . . . Come to think of it: What might that huge portion be? What's left for me to account for? We will find out on p. 170.

*See further tips on how to reduce your footprint for mobility and consumption of goods in part **4.** Solutions on pages 194 and 198.*

NORTHERN EUROPE IN 2017 . . .

... AND IN 2018

Left page: July 19, 2017
Right page: July 24, 2018

In 2018, an unusually long heatwave, lasting for several months, turned large parts of Europe brown. A high-pressure system had "got stuck" over the continent—result of the weakening Northern polar jetstream (see p. 154) due to changing climatic conditions.

Images taken by the Visible Infrared Imaging Radiometer Suite (VIIRS) on the NOAA-NASA satellite *Suomi NPP*. Source: NASA Earth Observatory

Here in the woods I feel connected to nature, which supplies me with so much I take for granted but actually need dearly to stay alive. Photo: Christian Schienerl

A WALK IN THE WOODS

Biodiversity, resilience, and ecosystem services

Before I head back to the city, I make use of the last winter light to take a walk in the woods close to the printers. I leave my stuff in the car and start out on foot, my every breath producing a large plume of water vapor in the cold air. It feels good to move my legs a bit after having been sitting and standing all day. The woods are very quiet without birdsong, and the snow along the path dampens any sound even further. As I stumble along the snowy path, I think back to the woodfire that was warming us at the office this morning. I could have collected wood for that fire from this forest. It will grow new volumes of it over the course of the next year. The leaves from the trees now rotting beneath the snow cover will turn into new humus, which in turn will feed the trees with new nutrients. Next year's leaves will sequester more carbon dioxide from the atmosphere (some of which I am exhaling at the moment with those large plumes of water vapor), and the trees will emit oxygen instead, which I need for breathing.

Here in the woods I feel connected to nature, which supplies me with so much I take for granted but need to stay alive: the air I breathe, the wood I burn or use for building things, the crops that feed me, the trees that shade me, the water I drink. I wonder what I am giving back to nature for all these ecosystem services she provides for free? (Ecosystem services are the contributions and benefits of ecosystems to economic and other human activity. An ecosystem is "a community of organisms and their physical environment interacting as an ecological unit."[58]) But of course nature does not ask for anything in return, indeed she neither "gives" nor "receives"—she is not there for me at all. The romantic notion that "Mother Earth" is providing for her human (and other) children is just the flipside to man's self-empowerment to exploit nature for his purposes. Both ideas are nothing but all too human moral tales. Nature evolved over almost 4.6 billion years, and has always worked exclusively with whatever was there at a given moment. Nature does not care about who believes what they may be entitled to.

Before life existed on Earth, the natural world was one of chemical elements and reactions among them—a world of rocks, minerals, water, gases, electricity, heat, and pressure.[59] When anorganic turned into organic chemistry by chance, and microbial single-celled life came into being, nature expanded to become a biological world.[60] Single-celled organisms stumbled upon ways to use whatever raw materials they found in their enviroment to ensure their own growth, survival, and proliferation. The twin evolutionary lessons of random mutation and natural selection "taught" them how to harness sunlight and other abundant materials in their world. Thus they became the first "masters of the universe." Never mind so many of the tiny creatures produced so much waste at some point—a waste-product called oxygen—that the whole planet began to rust and turn red. When there was no more stuff left to rust, the superfluous oxygen began to accumulate in the atmosphere, and over hundreds of millions of years it would change the entire planet again. Not only did ever larger-bodied lifeforms like plants and animals become possible with all that oxygen in the atmosphere—this newly developing *biosphere* would also react with the rocks on Earth, her *lithosphere*, and create an abundance of new minerals in a coevolutionary process, as Robert Hazen has powerfully demonstrated.[61] Over hundreds of millions of years more, all of Earth's spheres would coevolve ever more closely linked, supporting myriads of newly evolving lifeforms on the planet's surface, in its oceans, even inside its crust.[62] Over time, Earth's biosphere would permeate all of her spheres so profoundly that she herself became a living planet.[63] Just think of the natural carbon cycle (pp. 52–55) as an example of the many self-regulating and life-enabling flows of materials through Earth's spheres. An immense diversity of species, of individuals within species, and of ecosystems being inhabited and/or created by those species arose. Food webs from the tiniest of creatures—microbes—to the largest of mammals established themselves. Earth became a "complex and self-regulating system, in which everything is connected to everything else,"[64] as Johan Rockström describes it, alluding to Alexander von Humboldt's insights from the early nineteenth century (see p. 51). Rockström goes on: "Oceans, land, water, and biodiversity, through flows and stocks of energy, nutrients, carbon, and other elements" could regulate the persistence of a certain state of the whole Earth System—or trigger changes into wholly different system states. This property of a system (be it an individual, a forest, a city, an economy, or the whole planet) to remain within a certain system state and go on thriving even when it experiences sudden, external pressures or blows is called its **resilience**. Nature's resilience rests on several pillars, but **biodiversity** is one of the central ones. Earth experienced several blows to the whole system—catastrophes of global scale—during her long history. Some of them seriously weakened her resilience: At least five so-called **extinction events (EEs)** decimated biodiversity to a critical level, with numerous species being lost forever. The extinction of the dinosaurs 66 million years ago was just the most recent and prominent of the *Big Five* EEs.[65] But each and every time, over millions of years, new biodiversity would arise, strengthening again the resilience of the biosphere and of Earth as a whole.

The evolutionary principle of random mutation, which drives biodiversity, is simple and effective: Ever more diverse species adapt ever more perfectly to ecological niches and subsystems, spawning a rich web of natural possibilities with numerous redundancies. When a hit to the system occurs and environmental conditions change abruptly, a certain number of species will either already have adapted to these or similar conditions or be flexible enough to adapt quickly. Other species, unable to adapt rapidly enough, will perish.

This principle can be seen at work in the large number of different natural crop seeds, for example. Hundreds of differently adapted species of the same crop may exist: Some have evolved to be more drought-resistant, others to be more flood-resistant. When disaster strikes and a drought devastates the land,

WE ARE CURRENTLY EXPERIENCING THE sixth mass extinction IN EARTH'S LONG HISTORY. THIS TIME THOUGH IT'S *NOT AN* ASTEROID THAT HAS HIT THE PLANET . . .

↑ **The wellhead disaster at the "Deep-water Horizon" drilling platform** in the Gulf of Mexico on April 21, 2010, became history's biggest oil spill. Over months, hundreds of millions of liters of oil leaked into the ocean. Photo: U.S. Coast Guard

↗ **Adult brown pelicans covered in oil** from the "Deepwater Horizon" spill. The birds are waiting in a holding pen to be cleaned by volunteers at a bird rescue center in Louisiana, 2010.

Photo: Daniel Beltrá / Greenpeace

→ Due to **industrial overfishing** for decades, many fish stocks all around the globe have been depleted to levels at which the collapse of whole populations has become an imminent danger.

Photo: iStock.com / Paolo Cipriani

Cleared rainforest land for a mine outside of Munguba, Brazil. Photo: Daniel Beltrá / Greenpeace

Fighting forest fires laid for deforestation purposes in Sumatra, Indonesia. Smoke rising from smouldering peat-land in PT Raja Garuda Mas Sejati.
Photo: Ulet Ifansasti / Greenpeace

the more drought-resistant species will survive, still produce crops, and support the food web they are part of. The flood-resistant species may perish in the drought, but the whole system's resilience was based on its large biodiversity having brought forth drought-resistant crops among many others. This naturally evolved principle is completely turned on its head, when today's large agrobusiness corporations (like Bayer) artificially breed highly specialized seed species. They entice farmers in flood-prone regions to use only their artificial, highly flood-resistant seeds. But the longer these seeds are used exclusively, the more natural biodiversity in these regions is lost, and "[r]eductions in the diversity of cultivated crops . . . mean that agroecosystems are less resilient against future climate change, pests and pathogens."[66] When a different kind of disaster strikes—which happens more regularly with the climate crisis—agriculture may just not withstand that system stress, and collapse.

What is true for crop seeds can now be seen all across the biosphere. That next big hit to the system is not imminent—it has already struck the planet: We are currently experiencing the sixth mass extinction in Earth's long history. Between 1970 and 2014, populations of wild vertebrate animals have, on average, declined by 60%.[67] Populations of terrestrial insects show a similar decline: A recent meta-study finds "an average loss of 8.81% per decade in terrestrial ecosystems. Such a decline is concerning given the critical role that insects play in food webs and ecosystem services and may contribute to other changes such as the declines observed for some insectivorous bird populations."[68]

In 2019, the Intergovernmental Science-Policy Platform on Biodiversity and Ecosystem Services (IPBES) released its first assessment report on the status of biodiversity and ecosystem services on Earth. Its findings are disturbing: "An average of around 25 per cent of species in assessed animal and plant groups are threatened . . . around 1 million species already face extinction, many within decades . . . the global rate of species extinction . . . is already at least tens to hundreds of times higher than it has averaged over the past 10 million years."[69]

This time the forcing factor for biodiversity loss does not come from without the system but from within, from the one species having now overwhelmed the great forces of nature. "The biosphere, upon which humanity as a whole depends, is being altered to an unparalleled degree across all spatial scales. Biodiversity—the diversity within species, between species and of ecosystems—is declining faster than at any time in human history."[70]

This is bad news, as mankind and nature are now so strongly coupled that they form *one* socioecological system: "[I]n our globalized society, there are virtually no ecosystems that are not shaped by people and no people without the need for ecosystems and the services they provide."[71]

The major driving force behind biodiversity loss is the expansion of lands used for agriculture, mining, logging, and building. Highly balanced ecosystems are disturbed and put under immense stress by such human activities. Many natural habitats are simply destroyed. Increasingly, this leads to even more pressing dangers for humanity: Zoonotic diseases (viral diseases transmitted from animals to humans) like SARS, MERS, Ebola, and the recent devastating COVID-19 pandemic have become more frequent in recent decades. With humans intruding into formerly untouched wildlife habitats, physical contact with wild animals increases. Something as banal as a transport road to a remote mining operation being driven through virgin rainforest may create new contact zones in which zoonotic cross-contagion can happen. Conversely, wild animals under increasing ecosystem stress may penetrate into suburban or urban areas for food or shelter. Live animal markets (so-called *wet markets)* are another source of potential cross-contagion. The recent SARS coronavirus 2, which rapidly spread across the whole globe, originated in bats. Many viruses do, because they do not affect these animals as they do humans and other terrestrial animals.[72] In the case of SARS-CoV-2, bats are believed to have transmitted the virus to pangolins or other animals, which were then offered as delicacies on the wet market in Wuhan, China. The first human was infected, and from there, the virus spread worldwide using humans and their globe-spanning transportation systems as *vectors* (carriers).

While in the case of zoonotic diseases it is the *interspecies barrier*, which humans themselves have made increasingly permeable for viruses, changing climatic conditions and human transportation systems are also shifting geographical barriers for vector-borne diseases and invasive species. The Asian Tiger mosquito, for example, was originally native to tropical and subtropical areas of Southeast Asia. It transmits viral pathogens like yellow fever, Dengue, Chikungunya, and Zika. Due to the expansion of tropical climate zones and because it uses international passenger and freight traffic, the Asian Tiger Mosquito is now also found in Southern Europe, the Eastern US, sub-Saharan Africa, and South America. Changing climatic conditions and high interconnectedness of world regions under the conditions of globalization also allow for easy migration of *invasive species* (like ragweed in Europe), and for the proliferation of pests like the bark beetle, which in Austria has been causing increasing damage to forests in recent years as it is now able to breed not once but two or three times a year in the warming climate.

↑ **Bees and other pollinators—** both natural and managed—are on the decline globally, particularly in North America and Europe. In recent winters, honeybee colony mortality in Europe has averaged around 20%. Photo: Bas Beentjes / Greenpeace

→ **Adult Bonobo** in a Bonobo sanctuary in the Democratic Republic of Congo. The Bonobo is one of mankind's closest relatives but threatened with extinction from destruction of its habitats.

Photo: Filip Verbelen / Greenpeace

Sumatran tiger in the Melbourne zoo. In Indonesia, forest destruction for palm oil production is pushing Sumatran tigers to the edge of extinction, with only 400 left in the wild.

Photo: Tom Jefferson / Greenpeace

WITHIN MY LIFETIME, populations of wild vertebrate animals have, on average, declined by 60% (!)

A polar bear in drifting and unconsoli-dated sea ice, off Cape Clay, Greenland. Polar bears cannot survive without sea ice, using it to raise their young, to travel, and as a platform for hunting seals. The species is threatened with extinction because global heating is causing its sea ice habitat to melt away rapidly.

Photo: Nick Cobbing / Greenpeace

Stuck in a traffic jam. I can't move on, I can't get out, I am wasting lifetime in my car, and at the same time, I am laying waste to my environment.

Photo: Jochen Tack / Alamy Stock Photo

STUCK IN A TRAFFIC JAM

Crossing planetary boundaries

As I head back from the printers to the city, first the orange glow of sunset gives way to blueish twilight and then the all-encompassing darkness of night. Ahead of me I now only see the red taillights of other cars on the highway and an ominous yellow glow in the distance. These are the lights of Vienna, reflected in the clouds, forming a vast bright dome sitting on top of the city. The closer I get to that dome, the denser the red taillights ahead of me get.

As I cross the city limits and ever more roads and cars feed onto the highway, things begin to slow down to a crawl. I have now become part of the daily evening traffic jam along the city highway. There is no way out of it, there is no way around it: The remaining five kilometers to my home will take me at least another 45 minutes. At first, the thousands of cars ahead of me still move along at walking pace. But the closer we get to the bridge crossing the Danube river, that walking pace turns into stop-and-go: drive a few meters, then bring the car to a complete halt for what seems to be an eternity. At this point, I would progress faster if I just got out of the car and walked home.

As I look around myself, peek into the vehicles closest to me, I realize almost everyone is sitting in their cars alone. Some drivers are talking—I guess over hands-free mobile devices to someone else and not to themselves—, others are nervously puffing cigarettes while texting or listening to music. As far ahead as I can see along the straight stretch of highway, I see only immobile red taillights. A gaze into the rear-vew mirror reveals the same picture, only with the red taillights replaced by white frontlights.

I don't want to know what kind of *particulate matter* cloud I am currently trapped in. Naively, I just trust the air-regulating system in my car to filter out most of it. An equally invisible and odorless gas exits the tailpipes of thousands of fossil fuel-combusting engines around me: carbon dioxide. From those tailpipes, tonnes of CO_2 currently rise up into the black nighttime atmosphere, where the gas will accumulate and heat up the planet just a little more, when tomorrow's sunlight will warm its surface again. What's even worse: Right now, in this traffic jam, this happens serving absolutely no purpose. On the contrary: The mobility we drivers promised ourselves has turned into absolute immobility. One might even feel caged, locked up in one's vehicle, trapped in between thousands of others. I can't move on, I can't get out, I am wasting lifetime in my car, and at the same time, I am laying waste to my environment.

My present situation makes me think of an influential body of scientific work gathered over the last ten years. In 2009, an interdisciplinary team of 29 scientists from all around the world introduced the novel concept of *planetary boundaries*.[73] For the first time in history, an attempt was made to identify the most important regulating cycles and stocks of biogeochemical elements within the **Earth System** that keep it in the stable system state we have known for the last 12,000 years (during the interglacial period of the Holocene). Efforts were made to quantify the limits within which these regulating cycles and stocks of matter will work the way we expect them to. The nine **planetary boundaries (PBs)** which were identified are: *the integrity of the biosphere, climate change, novel entities* (= chemical pollution)*, stratospheric ozone depletion, atmospheric aerosol loading, ocean acidification, biogeochemical flows of nitrogen (N) and phosphorus (P), freshwater use*, and *land-system change*. The purpose of identifying these nine core properties of the Earth System and quantifying at what limits they would stop functioning as expected was to "define a safe operating space for humanity." This *safe operating space* is represented in the graphic on the opposite page by the green center circle.

You can see the results from the major 2015 update on the planetary boundaries[74] visualized in the graphic →. Two of the nine PBs have already progressed far beyond the *zone of uncertainty* (yellow) and into the *high-risk zone* (orange): One of the two main indicators for *biosphere integrity,* the *background extinction rate* (E/MSY = extinctions per million species-years) has completely gone off the rails. In layman's terms: Never have so many species gone extinct in such a short time as right now. The other dangerously overdrawn PB are the *nitrogen and phosphorus cycles:* mainly originating in industrial agriculture, these two elements enter the hydrosphere and biosphere at such unprecedented rates that they literally change the chemistry of the planet. Two further planetary boundaries—*climate change* and *land-system change*—have progressed beyond the safe operating space (green) into the zone of uncertainty. The *ocean acidification* boundary is nearing the zone of uncertainty, and only *freshwater use* is still within the limits of a safe operating space (green). For two to three of the PBs—*novel entities* (chemical pollution), *atmospheric aerosol loading*, and the Biodiversity Intactness Index (BII, part of *biosphere integrity*)—not enough measurements and data could be gathered yet to arrive at quantitative results.

The planetary boundary of *stratospheric ozone depletion* is special in that it once had already progressed into the zone of uncertainty but later receded again into the green zone. Until the mid-1980s, the use of *hydrofluorocarbons* (HFCs) had seriously depleted the ozone layer, and an "ozone hole" opened up in the stratosphere, letting in life-harming solar UV radiation. The subsequent global ban of HFCs (Montreal protocol) was swiftly enacted all over the world and led to partial recovery of the ozone layer in the decades to follow. In 2019, the ozone hole was the smallest measured since the 1980s.

The global ban of ozone-damaging HFCs is often cited as an example of swift and successful international cooperation in tackling an environmental crisis. More such global cooperation would be needed in the face of today's pressing environmental issues, it is maintained. While there is certainly need for a lot of such intergovernmental collaboration, one should be cautious about comparing the two crises. The ban of HFCs was a small and quite manageable

THE NINE
planetary boundaries
SHOW WITHIN WHICH LIMITS EARTH CAN FUNCTION THE WAY WE EXPECT HER TO.

BII = Biodiversity Intactness Index
E/MSY = Extinctions per Million Species-Years
N = Nitrogen
P = Phosphorus

■ = below boundary (safe operating space)
■ = in zone of uncertainty (increasing risk)
■ = beyond zone of uncertainty (high risk)

Graphic: J. Lokrantz / Azote based on STEFFEN *et al.* (2015)
Source: Stockholm University

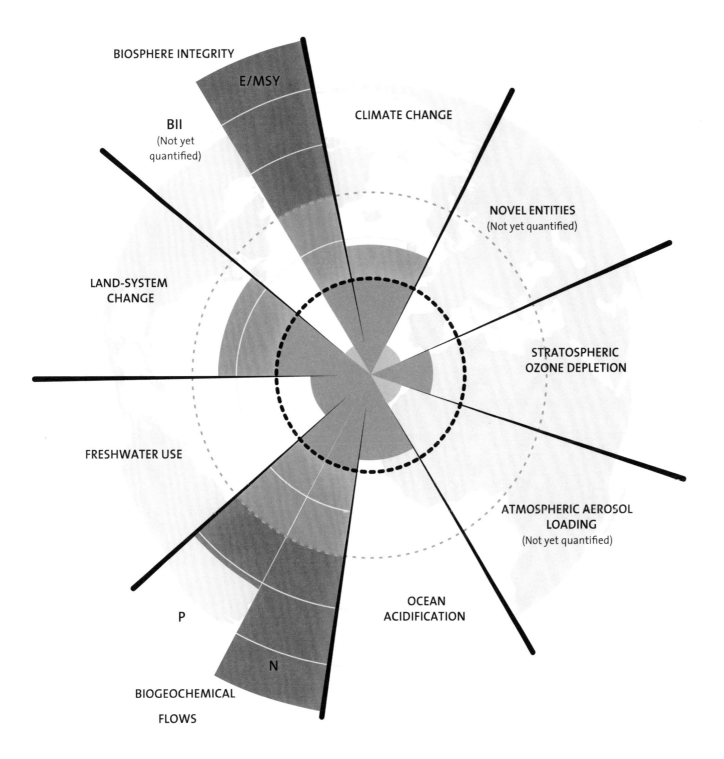

BIOSPHERE INTEGRITY

E/MSY

BII
(Not yet
quantified)

CLIMATE CHANGE

NOVEL ENTITIES
(Not yet quantified)

LAND-SYSTEM
CHANGE

STRATOSPHERIC
OZONE DEPLETION

FRESHWATER USE

ATMOSPHERIC AEROSOL
LOADING
(Not yet quantified)

P

N

OCEAN
ACIDIFICATION

BIOGEOCHEMICAL

FLOWS

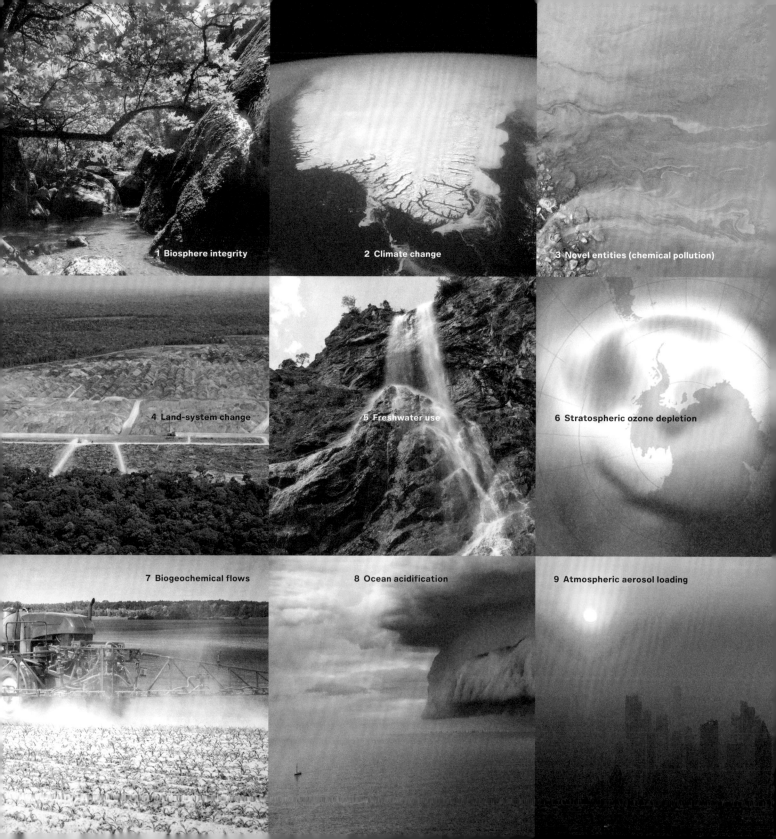

1 Biosphere integrity
2 Climate change
3 Novel entities (chemical pollution)
4 Land-system change
5 Freshwater use
6 Stratospheric ozone depletion
7 Biogeochemical flows
8 Ocean acidification
9 Atmospheric aerosol loading

feat compared to the scope and amount of harmful substances and greenhouse gases the international community has to deal with today.

As the chapter about fossil fuels (pp. 72–89) has shown, especially the emission of GHGs is an all-pervasive phenomenon in today's globalized economies and societies. What the world community has *pledged* to do to tackle the all-encompassing climate crisis we will see in the following chapter (pp. 156–169). But in regard to the target that the international community has set itself—limiting global heating to 1.5 °C over pre-industrial levels—what it has achieved so far amounts to nothing but failures and too-little-too-late. What *can be* done though with today's technologies and existing societal solutions—as discussed in the final chapter (pp. 188–205)—affords a positive outlook: Employing all means at our hands, we could turn around this story of failures into a success story, one of creating a more just and equitable world, in which economies serve societies and not the other way around—a world in which we operate within the boundaries of a highly-evolved, complex nature, whose elements all depend on each other.

As I take a look around right now at the numerous vehicles with their single passengers being stuck in this traffic jam, I wonder if there really still is a chance to achieve this. How do you get all these people to abandon their combustion engine cars and switch to public transportation or emission-free vehicles? Even provided this could be done, you would have to allow fewer vehicles onto highways, roads, city streets, and parking spaces, which occupy up to 50% of urban space. Nowadays everywhere I go and with almost everything I do, I get a feeling of there being too much of everything: too many people overcrowding spaces, too much consumption of goods, energy, and lifeforce. Even time seems to have become overfraught with too many things to do, too many places to be at the same time. It is obvious to me that collectively we are crossing numerous boundaries on local levels every day, and all these minor crossings all around the world translate almost directly into the crossing of the planetary boundaries discussed here. If we do not want to live in an entirely congested world, which cannot function any longer the way we expect it to, not taking the car to the printers is one important small step for man—which I resolve to take just now. But a great leap for mankind it isn't. That leap will have to come from far more pervasive solutions.

The emissions caused by my flight are a classic case of an "externality," where the costs of environmental pollution are simply dumped on nature.
Photo: Christian Schienerl

The Paris Agreement and tipping points

Around 5 p.m. I finally get off congested city roads and turn onto the street where I live. It's still early enough in the evening, and I quickly find a parking space close to my building. When leaving the printers, I had planned to swing by the office to pick up the new smartphone. But sitting in a hardly moving car for almost an hour and the outlook of not finding a parking space later on made me change my mind.

Everything else will also have to wait until tomorrow morning, as tonight I still have a special task ahead of me: I began the day dead-set on stopping global heating by this evening. So now I will do my final footprint accounting and find out how I have done. Unfortunately, there is one more thing I have to take care of before I can do so . . . and that thing is going to add to my Footprint substantially: I have to book a flight to the Canary Island of La Gomera for the beginning of March, when my best friend will marry there. Don't get me wrong: I look forward to the occasion and to finally visiting that island, which I have heard so many good things about. Also, La Gomera will be pleasantly warm in March, when temperatures here in central Europe might still be winterly, and we will have had five to six months of depressingly low temperatures, fog, snow (or mostly rain), and lack of sunlight. But **the five-and-a-half-hour flight** from Vienna to Tenerife **and back will cause almost exactly one tonne of carbon dioxide emissions.** That is far more CO_2 than I emit driving my car for private purposes all year (390 kg, see p. 136). It is also more than I emit heating my office all through that long winter (ca. 900 kg). So that one flight has quite a hefty pricetag attached to it.

Not so hefty if you look at what the airline will charge me: The cheapest non-stop return flight is 150 Euros. If we deduct the carbon price of 110 Euros per tonne of CO_2 emissions introduced earlier (p. 106) the airline would be left with a meager income of 40 Euros for the flight (a little more than today's 70-km-car ride to the printers cost me). But that is not how their ticket pricing works: 1. **The airline does not pay carbon tax**—so the 150 Euros will be all theirs. While inner-European aviation *is* part of the EU Emissions Trading System (see p. 82), airlines still receive the majority of their emission allowances for free. 2. Due to a tax exemption, **the airline does not pay tax on kerosene,** their aviation fuel. 3. **The airline is not required to add value-added tax (VAT) on the ticket price** in Austria (or anywhere else for that matter). What this boils down to: The airline does not pay anything for the CO_2 emissions incurred by that flight. Neither do I, or anyone else. So it's a classic case of the cost of environmental pollution simply being dumped on the environment as an economic "externality" (see pp. 105–106). That is the "polluter does not pay anything"-principle of today's linear economy in thrall to the idea of never-ending growth. This is the same economy that is so very fond of the belief that markets should and can regulate themselves. Well, both these ideas are nothing but neoliberal myths, as we had to learn the hard way over the past four decades. The very real prices of today's airline tickets are entirely based on these myths though (as are almost all other prices in this economy). But there is an enormous cost to living inside this fairytale and it is presented elsewhere now as a final overdue notice. The creditor is nature, and she will not be reasoned with. Either we pay up right away, or we will be put out of business—it's as simple as that.

This should not come as a surprise: **Warning signs and cautionary voices along the way were plentiful and present already from the end of the nineteenth century**. The heat-retaining properties of carbon dioxide (CO_2) and the resulting natural greenhouse effect have been well known since the work of Swedish physicist and chemist Svante Arrhenius, published in 1896.[75] In later works Arrhenius even predicted that increased burning of fossil fuels by humans was going to boost the greenhouse effect and would result in global warming.

In the early 1950s, scientists like Gilbert Plass spoke out on the fact "that the large increase in industrial activity during the present [twentieth] century was discharging so much carbon dioxide into the atmosphere that the average temperature was rising at the rate of 1 ½ deg. each century. [. . .] A]bout 2,000,000,000 tons of coal and oil were burnt each year, adding 6,000,000,000 tons of carbon dioxide to the atmosphere." (see newspaper clipping p. 37) These numbers from 1953 were astoundingly accurate (see graph p. 175)—but it wasn't until **1958**, when **Charles Keeling started measuring CO_2 concentrations in the atmosphere** at the Mauna Loa Observatory in Hawaii that their yearly increase could be verified and presented as the *Keeling curve* (see p. 61). Just a year earlier, the long-standing prior misconception that oceans would directly absorb the bulk of human-caused CO_2 emissions, had been proven to be wrong in a 1957 paper by Roger Revelle and Hans Suess.[76] Together with Keeling's measurements, that paper was the opening shot for global warming debates.

The Limits to Growth,[77] the first ever Report to the Club of Rome in 1972, drew on a simulation and prediction model for contemporary society developed at the Massachussetts Institute of Technology (MIT). Given the ongoing depletion of natural resources and increasing environmental destruction would continue, the software predicted that growth of the global industrial society would come to a grinding halt around 2030 and reverse into contraction.[78] Even fossil fuel companies like ExxonMobil conducted their own climate research in the late 1970s. Unsurprisingly, their research also clearly established a connection between the use of fossil fuels and the increase of atmospheric CO_2: "[T]he major contributors of CO_2 are the burning of fossil fuels [. . .] and oxidation of carbon stored in trees and soil humus . . . There is no doubt that increases in fossil fuel usage and decreases in forest cover are aggravating the potential problem of increased CO_2 in the atmosphere."[79]

Foreseeing that the problem would backfire on them in the future, **Exxon-Mobil and other major fossil fuel players** increasingly opted to obscure these findings during the 1980s, and instead **began investing heavily in disinformation**

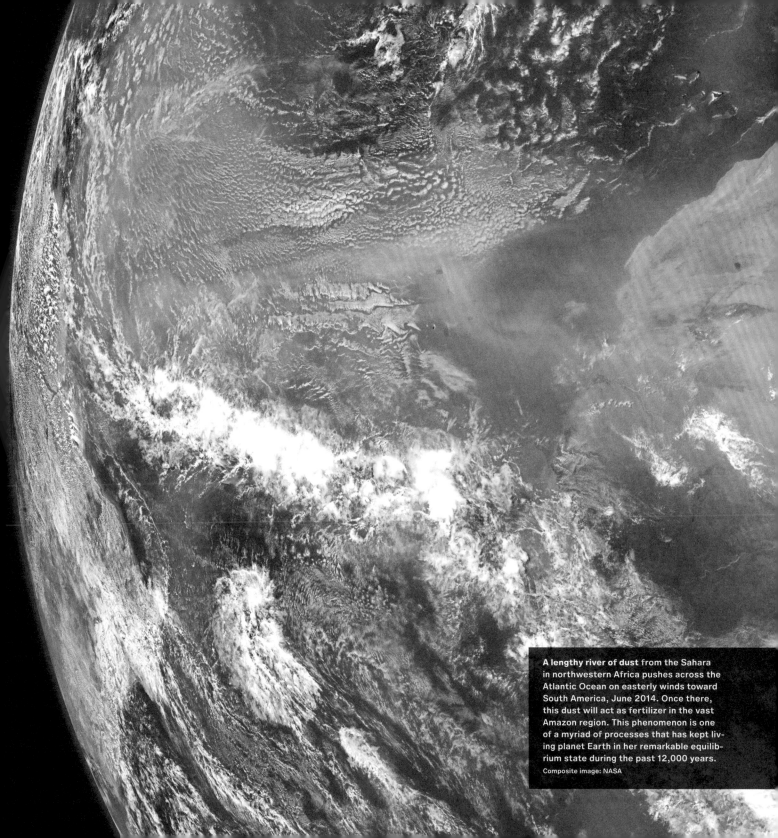

A lengthy river of dust from the Sahara in northwestern Africa pushes across the Atlantic Ocean on easterly winds toward South America, June 2014. Once there, this dust will act as fertilizer in the vast Amazon region. This phenomenon is one of a myriad of processes that has kept living planet Earth in her remarkable equilibrium state during the past 12,000 years.

Composite image: NASA

campaigning. They funded think tanks and political front groups which dissipated doubt about climate change, about the fact that it was man-made, about climate science, and even about the trustworthiness of science itself. In the US, organizations like the *Heartland Institute*, the *Cato Institute*, or *Americans for Prosperity* (all funded by fossil fuel corporations like ExxonMobil and Koch Industries) would employ their own "scientists" to professionally promote climate change skepticism, or even outright denial. In the US, this contributed to delayed political responses to the crisis in the 1990s. In 2018, six years after superstorm *Sandy* had devastated the Eastern US seaboard, New York and New Jersey attorneys general sued fossil fuel major ExxonMobil for fraud, maintaining that

The Amazon region, spanning eight countries and covering a third of South America, is often called the "lungs of the world." Vast areas of highly diverse and biologically productive virgin rainforest even create their own weather and precipitation patterns. This view—taken from the International Space Station while it was over the Brazilian state of Tocantins— captures a common scene from the wet season: Vast pillars of moisture rise via convection and then spread outward into anvil clouds as they collide with the stratosphere. Photo: NASA Visible Earth

... global greenhouse gas emissions have risen by 80%

DURING THE PAST THIRTY YEARS ALONE ...

the company had systematically lied to its investors about the risks of climate change during the 1980s.[80]

Despite climate change deniers' best efforts, by the late 1980s overwhelming scientific evidence for *anthropogenic* (human-caused) *global warming* had accumulated. In 1988 the United Nations therefore established the Intergovernmental Panel on Climate Change (IPCC), which was to regularly advise the governments of the UN on the current state of scientific findings on the matter. The IPCC has produced four major assessment reports to date. The next one is due in 2021.

The year 1992 saw the first UN Earth Summit in Rio de Janeiro, during which the United Nations Framework Convention on Climate Change (UNFCC) was opened for signature. The UNFCCC entered into force as an international environmental treaty in 1994. The Rio Summit had been preceded by a "World Scientists' Warning to Humanity," a declaration signed by 1,700 independent scientists from all over the world calling for a curtailment of environmental destruction and "a great change in our stewardship of the Earth and the life on it" if vast human misery was to be avoided. "A Second Notice" to the Warning was published in 2017, a "World Scientists' Warning of a Climate Emergency" in 2019.[81]

In the years following the Rio Summit some progress was made in tackling the deepening crisis, most notably with the adoption of the Kyoto Protocol in 1997. For the first time an international treaty committed the parties to it to a reduction in greenhouse gas emissions. The Protocol entered into force only in 2005, and currently 192 nations are parties to it. A similar treaty was meant to widen and deepen the Kyoto Protocol's engagements at COP15 in Copenhagen in 2009. The target of limiting global heating to 2 °C above pre-industrial levels was first introduced then. But the summit failed at adopting such an agreement. 2009 was also an exceptional year in that for the first time since the Rio Summit in 1992 global GHG emissions receded in the aftermath of the financial crisis of 2007/2008 (see graph p. 175). Unfortunately, 2009 remained a singular exception in that regard: Global greenhouse gas emissions have steadily risen since 1992—by a staggering 80% (!). The 2020 SARS-CoV-2 pandemic will probably lead to another marked dent in the global emissions curve. Ideally, this slump would be used as an impulse for a continuous bending of the curve henceforth.

In 2013, the IPCC released its Fourth Assessment Report (SAR5). This review of all scientific work on the subject contained so much irrefutable evidence of rapidly progressing human-caused global heating (even exceeding the IPCC's previous projections) that it became the cornerstone of the groundbreaking Paris Agreement, which was adopted by the world community in late 2015. The 189 signatory states acknowledged that "[t]his Agreement . . . aims to strenghten the global response to the threat of climate change" by "[h]olding the increase in the global average temperature to well below 2 °C above pre-industrial levels and pursuing efforts to limit the temperature increase to 1,5 °C above pre-industrial levels, recognizing that this would significantly reduce the risks and impacts of climate change . . ." *(Article 2*, Paris Agreement).[82]

After failing to make significant progress in combating climate change for the almost twenty years since the Kyoto Protocol, the Paris Agreement finally delivered an almost revolutionary leap forward. **Climate change was recognized as a global threat**, and ambitions to contain this threat were set surprisingly high. After all, 1 °C of human-caused global heating had already been reached by 2015, and limiting it to only 0.5 °C more would require major commitment to transformative change by all involved parties. The spirit of COP21 in Paris seemed to finally accurately capture the urgency of the climate crisis and the utmost importance of true societal change in the face of it. And there were good reasons for this sea change, as we will see further on. The revolutionary step taken with this agreement was that the world's nations were actually committing themselves to get the entire global economy off of fossil fuels—this process beginning by no later than 2020 and concluding by no later than 2050.

The next chapter will show how little actual commitment to this task has since matched the high spirits of Paris (pp. 172-185). But here and now, we have to look more closely at the very good reasons that informed the highly ambitious 1.5 °C goal. What could possibly happen if global heating was allowed to progress beyond that point?

We generally conceive of human-caused global heating and the resulting climate crisis as a simple, linear cause-effect relation: Humans put more greenhouse gases into the atmosphere (cause), and as a result global mean surface temperature will rise (effect). As a linear relation it is quantifiable: raise atmospheric CO_2 by X amount, and you will get Y amount of increased temperature. You can also reverse the equation and ask: If we do not want temperatures to rise by more than 1.5 °C above pre-industrial levels (the Paris target), how much more can we raise CO_2 levels in the atmosphere? That calculation leads to a **remaining carbon budget of 320 billion tonnes of CO_2 emissions** as of today before the temperature threshold of 1.5 °C will be reached and then be held for hundreds or thousands of years to come. Actually, the calculation supposes only a 66% chance for GMST to remain below 1.5 °C if the remaining budget will indeed be spent, and some experts believe that already today we have blown that chance. But let's stick to the prevailing carbon budget concept for now, and do a quick calculation: At current emission rates of almost 40 billion tonnes CO_2 per year, the remaining budget will be spent within eight years from now (in 2028/29). Eight more years to get a global economy soaked in fossil fuels entirely off of them! This is exactly what has to happen NOW. Because **if the 1.5 °C heating threshold is exceeded, mankind will enter an entirely new phase of climate disruption**. In this new phase our simplistic linear cause-effect relation (more human-caused emissions leading to

. . . SO THAT NOW—AT ANNUAL EMISSION RATES OF ALMOST 40 billion tonnes CO_2— *OUR CARBON BUDGET FOR STAYING BELOW 1.5 °C WILL BE SPENT WITHIN A FEW YEARS. AND THEN WHAT?*

↑ Smoke from man-made forest fires to clear land for farming, cattle grazing, and feed production, Amazon, Brazil.
Photo: Daniel Beltrá / Greenpeace

→ Amazon burning This satellite map shows the breath-taking devastation that took place in the Amazon region during summer and fall of 2019. Red and orange dots show forest fires burning, largely cases of arson. Image: NASA Visible Earth

Tipping point within warming range of:
1–3 °C 3–5 °C > 5 °C

Possible tipping cascades shown as

Arctic summer sea-ice

Arctic winter sea-ice

Permafrost

Boreal forest

Greenland Ice Sheet

Jetstream

Alpine glaciers

Indian summer monsoon

Thermohaline circulation

El Nino southern oscillation

Sahel

Amazon rainforest

Coral reefs

West Antarctic Ice Sheet

East Antarctic Ice Sheet

Graphic: SCHIENERL D/AD, based on STEFFEN *et al.* (2018), p. 8255; Photo: NASA

EARTH'S
tipping elements
CARRY THE RISK OF MANKIND LOSING ALL CONTROL OVER THE PROCESS OF GLOBAL HEATING.

a proportionate rise in temperature) is going to be suspended. Above 1.5–2 °C heating, positive reinforcing feedbacks in nature are going to kick in—the more, the higher temperatures get—and they are going to let temperature increases soar far beyond linear progression into exponential growth.[83]

The behavior of the ice masses on Earth (the *cryosphere)* provides an easy-to-grasp real-world example of what *positive reinforcing feedback* means: ice and snowpack are white and therefore reflect most of the sunlight hitting them. Therefore, white surfaces do not heat up as much as darker ones. Seawater on the other hand is a dark blue, and rock a dark brown or grey. Darker surfaces absorb more light. This property of surfaces to reflect more or less light is called *albedo*. Now, when ice starts to melt due to warmer temperatures (the linear cause-effect relation), it will expose more dark seawater or rock. The exposed darker surfaces will absorb more light and increase temperatures even more. Now ice is going to melt even more rapidly, in turn exposing even more darker surfaces. A positive reinforcing feedback has been triggered and speeds up former linear progression to exponential growth.

Earth has evolved into a highly complex system with intricately balanced elements in all of her spheres. Each of these elements contributes to the overall stable system state and depends on the stability of the others. You can disturb conditions (as humans have done), and a stable element is not going to change its state for some time—the duration depending on the element's resilience (see p. 142). You can disturb conditions even further (as humans do now) before anything novel will happen. But once the element reaches its threshold, it will abruptly switch to a different state and behavior.

Such thresholds are called tipping points. Natural elements with built-in tipping points are called tipping elements (see illustration p. 164). The linear changes they can undergo before they switch states are processes of accumulation—of building up to a certain point at which the full force of accumulated change will be released all at once. Think of a twig you bend into a bow: You can bend it for some time, applying force linearly. But at some point, the accumulated mechanical force will become too strong, and the twig is going to snap. It will abruptly change its system state.

"[T]here is growing scientific evidence that we may have reached a saturation point in terms of our pressures on Earth . . . The biggest global changes won't come from the triggers themselves, but rather from the positive feedback processes they unleash,"[84] warns Johan Rockström. Earth's tipping elements on p. 164 are color-coded for the degree of global heating that is going to push them to and beyond their tipping point. Unfortunately, this process itself does not follow a linear progression either but is characterized by "accelerating risks": The transition from 1.5 to 2 °C will result in proportionally and possibly exponentially higher risk levels than the transition from 1 to 1.5 °C, which we are currently seeing. "[T]his is a consequence of impacts accelerating as a function of distance from the optimal temperature for an organism or an ecosystem process."[85]

One tipping element we already witnessed reaching its tipping point at 1 °C global heating are the world's coral reefs. The oceans and marine ecosystems in them are particularly sensitive to changing temperatures now, as oceans have absorbed over 90% of the additional heat energy in the atmosphere caused by anthropogenic GHG emissions over the last few decades. Around 30% of carbon dioxide have even been absorbed directly by seawater, leading to increasing acidification of the oceans (see pp. 152–153). Under these fundamentally changed environmental conditions, in 2015 colorful coral reefs around the world appeared "healthy right up until the onset of mass coral bleaching and mortality, which can then destroy a reef within a few months."[86] (see image →)

Other tipping elements already seriously affected by 1 °C heating are Earth's large ice masses: alpine glaciers, the Greenland and West Antarctic Ice Sheets (both sitting on landmasses), and Arctic sea-ice. Late-summer sea-ice volume in the Arctic has decreased by nearly 80% since 1979.[87] Due to the amplifying feedback described above (albedo), the effects of global heating are felt two to three times stronger in the Arctic than anywhere else (this is called *polar* or *Arctic amplification).* The decline in ice has also already seriously affected the Northern polar jetstream (*Jetstreams* are large wavy windbands at high altitude which circumnavigate the globe and transport whole weather systems). Its oscillations extend further south now, and they slow down more often, which can lead to weather systems getting stuck in place (as was the case with the 2018 heatwave in Europe [see pp. 138–139]).[88] The increasing masses of ice melting into the seas from the Greenland and West Antarctic Ice Sheets, on the other hand, will not only lead to rising sea-levels but also change ocean temperatures and salinity. As a consequence, worldwide ocean circulation patterns like the thermohaline circulation in the Atlantic might also change. A Europe now constantly supplied with warm waters from the Carribean (Gulf stream) might paradoxically become much colder in an overheated world, in which thermohaline circulation in the Atlantic has reached its tipping point and may switch off.

2020 has already seen record-breaking temperatures not only in Western Antarctica but also in Arctic Siberia, accompanied by devastating wildfires in the boreal forests there. This presents the imminent danger of permafrost soils thawing and releasing vast amounts of the highly potent GHG methane. Permafrost thawing is considered to be one of the most dangerous tipping elements, because not only is methane 28 times more powerful as a GHG than CO_2 in the short run, the amounts of the gas frozen in Permafrost soils are so vast that their release would mean a gigantic leap in global heating.

The illustration on page 164 also features bidirectional arrows connecting some of the tipping elements. These arrows are there to show that once a tipping

→
Coral reefs and Arctic sea-ice
ARE THE FIRST TIPPING ELEMENTS WE HAVE SEEN REACHING THEIR THRESHOLDS AT 1 °C OF GLOBAL HEATING.

↑ **Coral bleaching around the Addu Atoll in the Maldives.** Algae living in the corals are released as a stress reaction caused by warmer water temperatures, thus turning the coral white. The corals can only revitalise once lower water temperatures return. If this does not happen, they die.
Photo: Uli Kunz / Greenpeace

← **Sea-ice melt ponds.** Crew from a U.S. Coast Guard vessel retrieve a canister dropped by a supply plane for NASA's ICESCAPE mission (Impacts of Climate on Ecosystems and Chemistry of the Arctic Pacific Environment). The canister was probably supposed to land on firm ice surface, 2011. Photo: Kathryn Hansen / NASA (CC BY 2.0)

Permafrost, Alaska
Coastal erosion reveals the extent of
ice-rich permafrost underlying the active
layer on the Arctic Coastal Plain in the
Teshekpuk Lake Special Area of the
National Petroleum Reserve, Alaska.

Photo: Brandt Meixell / U.S. Geological Survey

element (like Arctic summer sea-ice) has reached its tipping point, it will feed forward its changed state to other elements (like the Greenland Ice Sheet), and vice versa. More heating and less ice cover in either are going to amplify changes in the other as well. Once such interdependencies come into effect, the amount and speed of heating are no longer determined by the amount of anthropogenic greenhouse gas emissions. Once mankind has set off the first tipping elements, they will set off others by themselves. This process is called *cascading*. **Tipping cascades** cannot be predicted, and they will be entirely outside of human control and manageability. The risk of such cascades occurring increases with every additional fraction of a degree of global heating.

These are the good scientific reasons why the parties to the Paris Agreement set themselves the ambitious goal of limiting global heating to 1.5 °C. Above that threshold, mankind will quickly lose any and all control over how the process of global heating is going to progress. At any point in time from then on, the climate crisis can swiftly turn into a global climate catastrophe.

MY FOOTPRINT 4
– SERVICES (?)

The last time we met (p. 136), I wondered what resource uses I still have to take stock of after having already accounted for my energy use, my consumption of food and goods, and my mobility needs. So, what's still unaccounted for?

The answer to that question comes as a shock—even more so as it should have been evident from the beginning: Even before I had consumed anything today—in fact, while I was still sound asleep this morning—a 23%-portion of my Footprint called **"services"** had already been filled. What? How? I do not remember signing up for any such thing . . .

I ask Global Footprint Network what those ominous "services" might be, as they just add that quarter-portion to my tally without asking me anything else about my lifestyle. Here is their answer: "**A person's Ecological Footprint includes both personal and societal impacts**. The Footprint associated with food, mobility, and goods is easier for you to directly influence through lifestyle choices. However, a person's Footprint also includes societal impacts or 'services,' such as government assistance, roads and infrastructure, public services, and the country's military. All citizens are allocated their share of these societal impacts. This is why, if we want to achieve sustainability, we need to focus on both our own lifestyle as well as influencing our governments."[89]

Ok, let me get this straight: I live in a country that owns and maintains infrastructure (like roads, sewage plants, hospitals, schools, and military installations) and offers services to its citizens (policing, public schooling, universities, health services), whether they want or need to use these services or not. And ok, it seems fair to distribute the Ecological Footprint of those "services" to all citizens equally. Who else would you place that burden on? But this also means that I don't have any control over that 23%-portion of my personal EF. The only way for me to change something about that is to influence or pressure my government and public authorities into getting those services' Footprints to a level compatible with the resources of one Earth. That sets me back quite a bit in my endeavor to stop global heating on my own by tonight.

The final stock-taking of my personal use of resources has now clearly shown what I had already feared: Despite my best intentions, and notwithstanding the fact that I actually achieved significant cuts in all areas of consumption, **my Ecological Footprint remains almost twice the size of a sustainable one**—my levels of consumption require 3 global hectares, or the resources of 1.8 Earths, footprint-calculator.org tells me.

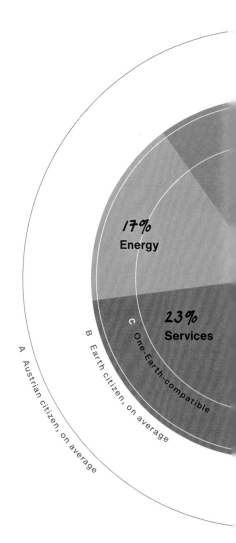

17%
Energy

23%
Services

A Austrian citizen, on average

B Earth citizen, on average

C One-Earth-compatible

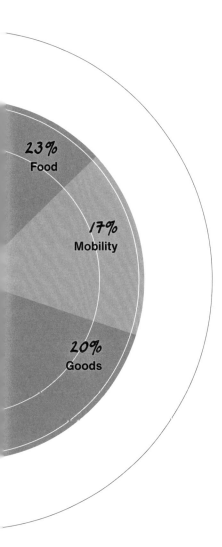

23%
Food

17%
Mobility

20%
Goods

So as of six o'clock this evening, I have to admit that I failed to deliver on my promise: I could not stop global heating. My Footprint still devours the resources of almost two Earths. I will not live sustainably from tomorrow on.

But why? Sure, I could drive my car even less than I do now, or get rid of it all together. I could completely refrain from flying. I could buy and make do with even less stuff than I have now. I could move to a smaller apartment or heat it less. I could eat less or less carbon-intensive food. Taking really radical stances in all these areas, I could probably shrink my Footprint to become one-Earth-compatible, even in the high-income country I live in with its high-consumer lifestyles and the societal services it provides to me and all other citizens. But to achieve that I would actually have to shrink my self (which is the literal premise of the film satire *Downsizing)*. Yet shrinking my self to leave a one-Earth-compatible Footprint would make me "fall out of place." My behavior and lifestyle would not fit into my socioeconomic environment anymore. I would become "a stranger in a strange land."

Since I don't want that, I have to pinpoint the problem elsewhere. If I cannot get to a sustainable lifestyle just by making conscious choices about what and how much of it I consume (energy, products, services, food, space), then maybe excessive resource use was embodied in all these things even before I touched and used them? There might be some truth to a suspicion I have had for some time now: *You cannot live sustainably in an unsustainable world* (to paraphrase a famous statement by German philosopher Theodor W. Adorno: "There is no right life in the wrong one.")

I will have to learn more about this dependency of individual lifestyle choices on the socioeconomic and political system they are made in in the following chapter. Maybe then I will see what else I could do to reduce my resource over-dependence.

To see how resource security for all might be achieved, see part 4. Solutions *on pages* 198 *and* 203.

Earth Strike
Sept. 27, 2019, Vienna, Austria
Photo: Christian Schienerl

WHY COULDN'T I STOP GLOBAL HEATING?

NOW is the decisive moment

It's 6 p.m.—As I had to learn over the course of this day, the globalized world we live in NOW—based on an "extractivist" linear economy and its take—make—waste approach opposite nature—has become unsustainable by the one planet we all share. The problem does not primarily lie with high levels of individual consumption though (the question of *how much*) but rather with what socioeconomic and political systems (markets of goods and ideas) offer in terms of goods and services for consumption (the question of *what):* Making a choice between unsustainably produced goods and ones that are just a little less unsustainable really is no choice at all.

I tried to do *my part* as a consumer today, yet my efforts have only taken me half the way, and I wasn't able to stop global heating on my own. So here is my new resolution for tomorrow: I will begin to fight for change on a systemic level. I will join the young (and older) protesters of Fridays for Future, Extinction Rebellion, and the Sunrise Movement. We will protest, strike, and petition our governments on every possible matter so as to force politics into promoting sustainable consumer and industry choices and sanction unsustainable ones. We will vote for ecology-conscious parties and fight against ecology-unconscious ones. We will join class-action lawsuits against companies exploiting and degrading our living environment. We will support strong regulative measures and laws on the national, EU, and international levels prohibiting corporations from making profits at the cost of destroying nature. We will support taxation that will make polluters pay for what they have done and continue to do.

That is going to be our fight from now on because we Earth citizens cannot rely any longer on free markets getting anything else done than making a few rich at the cost of the many. Unfortunately, we cannot rely on politics anymore either to make the right decisions for us. Rather it will have to be the other way around from now on: *We* have to make the right decisions and then force our elected leaders to implement them. All too often now, entanglement of parties and politicians with retro businesses (think of US president Trump and a large portion of the US Senate) and a seriously weakened state and public sphere (after forty years of having swallowed neoliberalism's bitter pills) are working against people's interests instead of for them (think of the coup-like Brexit referendum).

If Greta Thunberg asks political and economic leaders in Davos how they dare steal her future, the loud and clear answer is: profit, shareholder value, blind greed. It's not as if they hadn't known or didn't believe their own climate science (think of ExxonMobil, see pp. 158–160)—it seems they just didn't and still don't care.

When it comes to international cooperation in averting disastrous outcomes of the climate crisis, the same stubborn self-interest is at work on the level of nations pitted against each other in global competition for wealth and power. As groundbreaking as the Paris Agreement was in setting the highly ambitious goal of limiting global heating to 1.5–2 °C, as flawed it was in determining how this goal should be achieved. All it offered in that respect was that nations should "undertake and communicate ambitious efforts" as their "nationally determined contributions [NDCs] to the global response to climate change" *(Article 3)*. In contrast to true emission reduction *commitments*, which would have been binding under the terms of a treaty, the notion of voluntary, nationally determined *contributions* had been forged ahead of the Paris conference. It was supposed to overcome opposition from several nations (including the US, China, and India) against a treaty with legally binding emission reduction targets for individual countries. In a way, NDCs were what enabled the high-spirited agreement in Paris in the first place. We will get back to them shortly.

But let's first look at what the Paris target means in terms of required emission reductions.[90] Regard the graph on page 175 →: It shows the measured rise of global CO_2 emissions from the 1950s until NOW and what **a steep and rapid decline is needed from 2020 onward to achieve a gradual phase-out of fossil fuel emissions, as required to meet the Paris goal of limiting global heating to 1.5–2 °C.** After global emissions should have peaked by NOW, they need to be halved (!) by 2030 from currently almost 40 billion tonnes to 20 billion tonnes annually (equaling levels during the mid-1980s). Such a sharp decline means a reduction by almost 10% per year (!). Nonetheless, this first phase can be considered the easiest as all the "low-hanging fruit" can be picked first. But then emissions would have to be halved again to 10 billion tonnes annually by 2040 (equaling levels of the early 1960s). This second phase will be so much harder than the first. Yet between 2040 and 2050 emissions would need to be halved again to 5 billion tonnes annually (equaling levels of the early 1950s). Again, this phase will be so much harder than the previous. The remaining 5 billion tonnes of CO_2 per year would be considered *net-zero emissions* as artificial and natural carbon sinks (mainly vegetation) should be able to sequester and store them.

The emission reduction pathway delineated here is sometimes called the new **Carbon Law** (halving emissions every decade). Preferably, it should be made into nationally and internationally binding law. Because let's be clear about the physical facts once more: **To stand a 66% chance of limiting global heating to 1.5 °C above pre-industrial levels, we may emit only 320 billion tonnes more of CO_2** (that is, if you apply a very optimistic calculation approach). Once that **carbon budget** is spent, global heating of at least 1.5 °C is predetermined for the next few thousand years, as CO_2 accumulates in the atmosphere and lingers there for millenia.

This calculation is based entirely on a simple cause-effect relation with linear progression though. It does not take into account the increasingly likely effects of positive reinforcing feedbacks in nature—the tipping elements we already witness nearing their respective tipping points at only +1 to 1.5 °C (see p. 166–169).

The Paris temperature target clearly acknowledged these latest findings of climate science back in 2015 and committed mankind to keeping the risks of

→

NOW, ONLY ONE QUESTION REMAINS: Can we bend that curve in time?

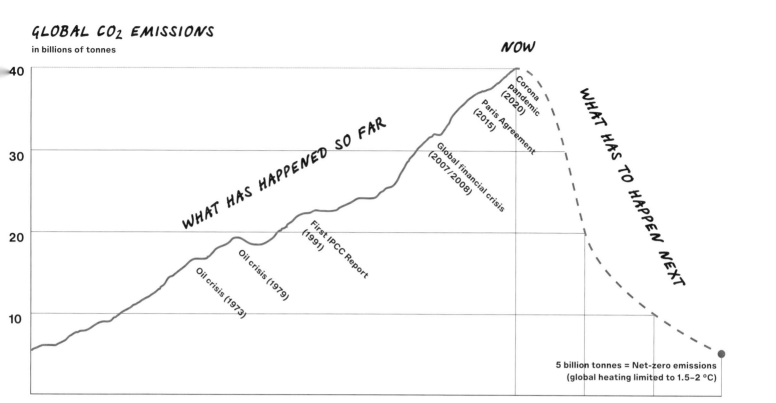

GLOBAL CO₂ EMISSIONS
in billions of tonnes

NOW

WHAT HAS HAPPENED SO FAR

WHAT HAS TO HAPPEN NEXT

Corona Pandemic (2020)

Paris Agreement (2015)

Global financial crisis (2007/2008)

First IPCC Report (1991)

Oil crisis (1979)

Oil crisis (1973)

5 billion tonnes = Net-zero emissions
(global heating limited to 1.5–2 °C)

40
30
20
10

1950 1960 1970 1980 1990 2000 2010 2020 2030 2040 2050

Data sources: U.S. Department of Energy,
2015 / The Global Carbon Project, 2018

reaching tipping points as low as possible. Since these risks are neither calculable nor manageable (mitigation and adaptation to climate disruption will be out of the question), the Paris Agreement recognized that the climate crisis could rapidly turn into a global natural disaster beyond human influence. A runaway greenhouse effect could even lead to **a *hothouse state* for Earth (with a higher than 4 °C global temperature increase)**. Such a hothouse state is the polar opposite of an ice age. It would certainly be equally devastating to human civilization, which could only be built during the stable climatic conditions of the "Goldilocks era" of the Holocene interglacial. The new Carbon Law (halving emissions every decade) is the one and only roadmap to achieving the Paris temperature target of 1.5–2 °C because it minimizes the risks of crossing tipping point thresholds.

So, how is the world community currently doing in following this Carbon Law and the resulting roadmap for emission reductions? In short: even less than poorly. The voluntary pledges or nationally determined contributions (**NDCs**) submitted to the UN by member states ahead of the Paris conference proved **highly insufficient to meet the 1.5–2 °C target** right from the beginning. Signatory

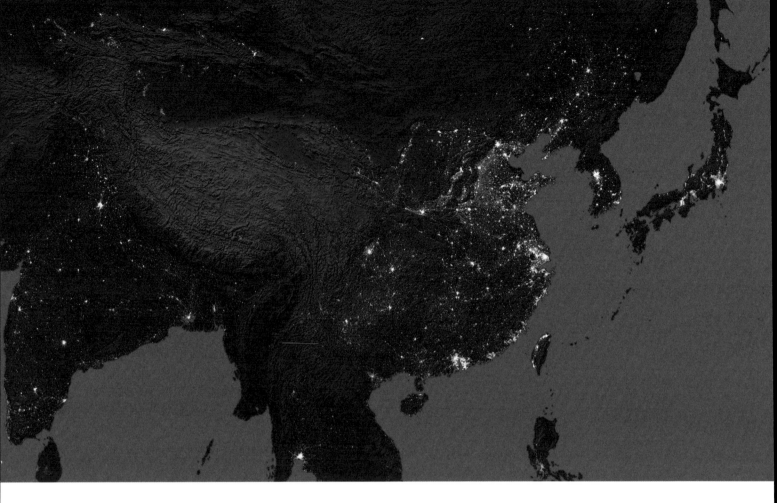

nations to the Agreement also acknowledged this *emissions gap*. But *Article 3* maintained that "[t]he efforts of all Parties will represent a progression over time," meaning that NDCs are supposed to be increased gradually, resubmitted and reviewed every five years. The first round of improved NDCs is due in 2020. In the fall of 2019 "70 countries announced their intention to submit enhanced NDCs in 2020, with 65 countries and major subnational economies committing to work towards achieving net zero emissions by 2050."[91] But with most of the G20 members[92] visibly absent, the likely impact on the emissions gap will be limited.[93] The economies of the G20 are responsible for 80% of global greenhouse gas emissions.[94]

There is still no mechanism to effectively sanction a state's failure to meet its own voluntary emission reduction pledges, much less force it to improve its NDCs. Most shockingly but perhaps unsurprisingly, not even the insufficient initial NDC pledges are being kept by most high-polluting countries to this day. At

This satellite map shows the **carbon emissions intensity** of different locations, especially cities and their hinterlands: the lighter the area, the more carbon is emitted. Bright parts of India (lower left) contrast with the dark area of the Himalayas and Myanmar right of them. Large areas of China in the center of the image are lit up brightly, especially in the regions surrounding Beijing and coastal cities Shanghai, Hong Kong, and Ghuangzou. Also clearly visible on the right: the cities of Seoul in South Korea and Tokyo, Kyoto, and Osaka in Japan.
Image: NASA Visible Earth

Massive change is coming

WHICHEVER PATH WE'LL FOLLOW. SO WE SHOULD ASK OURSELVES: DO WE WANT TO SHAPE THAT CHANGE OR NOT?

the R20 Austrian World Summit in Vienna in May 2019, UN Secretary General António Guterres summed up the dilemma in his speech: "We are doing less than what we already know would not be enough in relation to the engagements that were made in Paris."

Article 4 of the Paris Agreement stated that "[i]n order to achieve the long-term temperature goal set out in Article 2, Parties aim to reach global peaking of greenhouse gas emissions as soon as possible . . ." Obviously, again it had not been possible to set a clear deadline for **peak emissions** in the Agreement text. Instead, a nebulous "as soon as possible" was the go-to formula. Consequently, global greenhouse gas emissions are still far from peaking: instead they have risen at a rate of +1.5% per year during the last decade. The 2020 SARS-CoV-2 pandemic will put a dent in that curve but whether it can create enough momentum for a sustained downward trend is an entirely different question.

So, all rethoric and wishful thinking aside, what is the sober, no-nonsense outlook? **With current real-world emissions** *(business as usual* or *current policies* scenario) **mankind is on a path to an at least 3 °C hotter Earth** by the end of the century (possibly even 4.1 °C).[95] This equals the *hothouse state* scenario mentioned before. Most experts agree that human civilization as we know it cannot exist in such an overheated world.[96] **If current pledges (NDCs) were to be kept** *(pledges & targets scenario)*, **the world would still be on a path to a 2.8 °C hotter world.**[97] To make that outlook even more somber: Neither of these two scenarios factors in tipping elements and possible tipping cascades, which will become highly likely at heating beyond 2 °C.

So, as things stand, and without overstating the facts: Currently, we are collectively on a path that will end the world as we know it, and bring about a new one that we are ill-prepared to face. Therefore, **massive change is going to happen during the next few decades, whichever path we will follow and whether we like it or not.** But we still have a window of opportunity to determine what we want that massive change to look like. That window is going to remain open only for a few more precious years though.

There is no doubt: Implementation of the Carbon Law (halving emissions every decade) will require Herculean, internationally concerted efforts from all involved parties (and that is: *all* parties). Nations, regions, cities, communities, industry, businesses from small to transnational in size, the finance sector, and every single Earth citizen will have to contribute their fair share for this to work.

But **responsibility and ability to contribute to this task are not evenly distributed around the globe.** When we looked at how much CO_2 has been emitted in total globally between 1751 and 2017 (pp. 86–87), we found that only 30 of the world's 197 countries are responsible for 60% of its cumulative emissions: the USA alone for 25%, the now 27 EU countries plus the UK for 22%, and China for 12.7%. If we look at current emissions, the 20 member states of the G20 account for roughly 80% of global GHG emissions.[98] Pretty much the same picture emerges if you look at global GDP (gross domestic product) as an indicator of wealth and levels of consumption: The 30 countries of the OECD (Organization for Economic Cooperation and Development) accounted for 74% of global GDP but only 18% of the world population in 2010.[99]

Various terms have been used in the past to distinguish these 20–30 wealthiest nations from the other countries of the world. We will use the terms Global North and Global South here, primarily because they make reference to the climate and other of today's environmental and economic crises as global phenomena: In principal, the impacts of these crises are felt the same everywhere around the world, regardless of where they have been caused primarily. If that were not unjust enough, it is not even true to say that impacts are felt the same everywhere. Low-income countries of the Global South are much more vulnerable to impacts of climate disruption than are the high-income countries of the Global North, where mitigation and adaptation measures can be financed much more easily. The fate of island nations in the Pacific like Tuvalu and Kiribati illustrate this fact most emphatically: having contributed next to nothing to anthropogenic global heating, they are the most impacted by the climate crisis right NOW as they face total loss of their ancestral lands to the sea due to rising sea-levels. Indigenous peoples in the rainforest regions of the world, who have been stewards of these unique ecosystems for centuries, are also among the *frontline communities* suffering the most from causes and impacts of the climate crisis.

The unequal distribution of responsibility for and impactedness by climate disruption is an issue of equity and climate justice. It does not only concern different communities, groups of countries, or regions of the world but also different generations. Growing up in an increasingly wealthy country (Austria) after WWII, my parents and their generation perhaps profited the most from the socio-economic growth and material welfare brought about by the Great Acceleration (pp. 129–131). That wealth trickled down to the generation of my brother and myself. Growing up in the 1970s to 1980s, we still enjoyed a marvelous natural world. From the alpine glaciers of Austria and Switzerland to the rainforests of South America and wildlife in the Serengeti, I have witnessed the beauty and abundance of our home planet with my own eyes. But while the wealth of previous generations still continues to trickle down, the generation of my nieces now already faces an intensely altered natural world: Degraded by extractivism and overburdened by human economy and pollution, the Earth they get to see is a beaten and oppressed version of the one I saw. During my lifetime alone, populations of wild vertebrate animals have, on average, declined by 60% (!). Deprived of the opportunities I still had, Greta Thunberg and the youth of today rightly demand: "What do we want?—CLIMATE JUSTICE!—When do we want it?—NOW!"

→

Extreme weather events— THE FIRST DRASTIC SYMPTOMS OF CLIMATE DISRUPTION—HAVE BECOME MORE FREQUENT AND INTENSE DURING THE PAST 20 YEARS.

Extreme weather events such as

↖ **heatwaves** (a dry European city park during the 2018 heatwave. Photo: Mark Ramsay, CC BY 2.0),

↑ **droughts** (men drawing water from an earth well during one of the worst droughts in Mali, 2010. Photo: Vello Coviello / imaggeo.egu.eu, CC BY-SA 3.0),

← **superstorms and storm surges** (flooded New Orleans after Hurricane *Katrina*, 2005. Photo: David Mark / pixabay.com, CC 1.0),

179

↑ **wildfires** (California, 2018.
Photo: Adobe Stock / neillockhart),

→ **deluges, floods, and mudslides**
(Peru, 2013. Photo: Alberto Orbegoso,
CC BY 2.0)

have become measurably more frequent
and intense over the past couple of
decades. Extreme weather events which
occured once in a century in the past,
may now occur once in a decade.

The time to ACT is NOW!

MID-TERM GOALS FOR 2030 AND 2050 WILL BE POINTLESS, IF WE HAVE LONG SPENT THE REMAINING CARBON BUDGET BY THEN.

And NOW is exactly when implementation of the Carbon Law has to happen. **We have literally run out of time to bend that emission curve** (p. 175). During the past ten years, "[c]ountries collectively failed to stop the growth in global GHG emissions, meaning that deeper and faster cuts are now required."[100] Even deeper and faster cuts are required from the countries most responsible for cumulative emissions over the past 250 years, so that emerging economies get a chance to also prepare for the transition.

So, it's all well and good that the European Parliament declared a climate and environmental emergency in November 2019, and the newly appointed European Commission proposed a **Green Deal**[101] and a European Climate Law[102] that are supposed to make the continent climate-neutral by 2050. But critics—including 12 member states of the EU—say that concrete emission reduction targets for 2030 (50% compared to 1990 levels) are not ambitious and bold enough to meet the urgency of the situation. At least a 60 or even 65% reduction in EU-wide carbon emissions would be needed to stay in line with the Paris Agreement targets. Whether it should be 50 or 65% by 2030 is not really the question though (in reality the number would have to be closer to 75% actually): **If we do not take decisive and forcible action right NOW** and cut emissions by 10% by the end of this year (and go on doing so during each year to come), **the mid-term goals for 2030 and 2050 will be pointless**. The remaining carbon budget will be long spent by 2030, and global heating will definitely reach 2 °C and beyond, before we even start to seriously take action. **The time to ACT really is NOW!** Debates about this have lasted for over thirty years, and that is at least twenty years too long: If serious steps toward emission reductions and a decarbonization of the world economy would have been taken around the time of the Kyoto Protocol in the mid-1990s, reduction pathways would have been much gentler and easier to follow. **NOW what's really needed is immediate and disruptive change.** We could learn a few valuable lessons from the 2020 coronavirus pandemic in that regard: Financing a global economic bailout worth trillions of US-dollars, Euros, and Yuan does not seem to be impossible.

There is more good news: Today, hardly anyone is fooled any longer by climate change deniers and protractors of the antiquated fossil economy, who try to squeeze the last petrodollar out of this world. During the last few years acceptance of the facts of anthropogenic global heating and concern about them has become widespread[103]—not least due to the activism of Fridays for Future and other youth movements. It has also become obvious that the means and tools for the required fundamental transition are all there. They are readily available, they offer viable economic alternatives, and growing numbers of Earth citizens are willing to see them deployed immediately and at scale (see pp. 186–205). Another hopeful development is taking place in the powerful financial sector: Out of fear that investments in fossil fuel industries might soon end up as so-called "stranded assets" during transition to a decarbonized economy, more and more investors (among them the large pension funds) *divest* from these industries and reroute monies to green infrastructure investments.[104]

But still: achieving a transformation at the required scale is an unprecedented challenge for mankind. If you look at what humanity has achieved over the last 200 years though, and considering that the climate crisis is entirely man-made, you can find hope again: this crisis can also be resolved by mankind. No magic trick, screwball techno-fix (like *geo-engineering),* or divine intervention are required. It will take a lot of will, ambition, effort, and hard work. But it can be done— by us, for us. And that "us" includes everyone alive today, future generations, as well as all our fellow species, which suffer tremendously from our past transgressions but can do absolutely nothing about them. The 2020 coronavirus pandemic has shown what kind of societal forces can be mobilized at a time of crisis and to what extent funds can be freed up to finance necessary measures. Economic rebuilding after the pandemic should direct funds and economic stimulus packages to support exclusively the required green transition.

We have recognized that the world we live in today has become *unsustainable* by our one home planet—increasingly we now recognize that this world is not *inevitable* either. What's more, there are great opportunities to achieve change for the better in a lot of respects through the necessary transition: Why not make it a more just and equal world, while we are at it? One in which people, their well-being, and dependence on the natural world are at the center of attention, and not shareholder value, bank bailouts, and the financialization of all spheres of life. The European Green Deal aims to do just these things.[105] But intentions alone do not suffice anymore. Now that we are clear *where* we want to go, let's finally get going! The (non-)alternative—the world we are headed for if we stay on the current course of "eternal economic growth" and utter dependency on fossil fuels—will be one completely changed by unmitigated climate disruption. That world is as sure to follow from continued inaction as it will be devoid of *any* positive outcome. The displacement of hundreds of millions of people who will have to abandon their homes because of climate disruption alone would completely destabilize social, economic, and political systems worldwide. Not to speak of the mounting costs for the damage brought about by ever more frequent extreme weather events and resulting economic shutdowns.

So, while we now all seem to be on the same page finally, the moment has come to take action and turn the page to a new, decarbonized, and just chapter in human endeavors.

→ Flying through smoke clouds
The most devastating bushfires ever were raging in Australia's southeastern New South Wales and Victoria provinces in late 2019 and early 2020. It is estimated that more than one billion birds, mammals, and reptiles, many unique to Australia, have been killed by these fires. Photo: NASA

↓ A Rural Fire Service volunteer rescues a possum with burns fleeing a fire on the outskirts of a small town in New South Wales. After blazes broke out in September 2019, Australia saw unprecedented bushfires destroying nearly 11 million hectares and at least 29 people losing their lives. Photo: Kiran Ridley / Greenpeace

Decommissioned oil rigs parked in the Cromarty Firth, Scotland, where they await transport to scrapyards. The oil downturn of 2014, when prices for crude oil dropped steeply, hit notoriously expensive North Sea oil production the hardest. Cheap US and Canadian oil from fracking (see p. 89) and tar sands (see p. 80) made those two major consumers less dependent on imports. Slowed-down growth in emerging economies China and India reduced demand even further.
Photo: iStock.com / grafxart8888

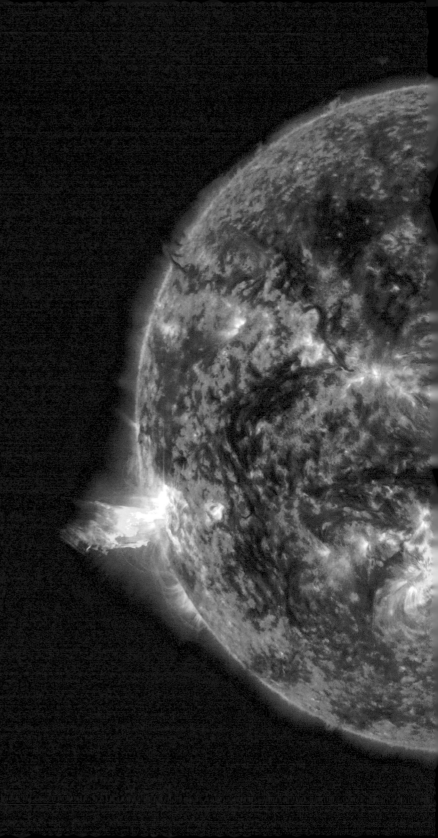

4. Solutions

A practical guide
to shrinking our
footprint on Earth

The enormous nuclear fusion reactor at the center of our solar system (the Sun) will go on fusing atoms for another five billion years. The amount of energy the Sun sends to Earth at any given moment is 10,000 times what mankind currently consumes in total. Photo: NASA

This final chapter looks ahead to tomorrow, and at existing solutions to mitigate the climate crisis in the fields of **1. energy use**, **2. mobility**, **3. industrial production**, **4. agriculture and land-use**, and **5. policies and regulation**. From page 189 onward, measures that each one of us can take individually to shrink our Ecological Footprints *(individual solutions)* are listed side by side with measures that have to be taken by governments on the international, national, and regional levels, by cities and communities, by industry and businesses small and large, by the financial sector, and by us Earth citizens collectively *(societal solutions)*. The presented remedies show that a sustainable world is not only *conceivable* but also entirely *possible to achieve* with the technological and intellectual means we already have at hand.

1. Energy

Most people on Earth live in **highly energy-dependent societies** today. Our economies, technologies, and everyday cultures run on vast amounts of readily available, cheap power. For the past 200 years fossil fuels have served that purpose and have driven enormous economic growth and gains in material welfare for a rapidly expanding world population. But fossil fuels were an invaluable and non-renewable energy subsidy from the deep past. Not too far in the future they will be depleted, and then that will be it. Long before fossil fuels could run out though, their burning has caused a far more profound predicament for mankind. Global heating and the concurrent climate crisis now pose an existential threat to everyone and everything that has been achieved. We know we must not go on using fossil fuels as energy sources, yet our lives seem to entirely depend on them: 86% of global primary energy demand is still met by coal, oil, and gas.

Since energy-dependency itself is not going to decline in the near future, we need to **substitute dirty fossil energy sources with clean, renewable sources**. That transition has already begun years ago, but it has to be sped up, scaled up, and intensified considerably during the next few (see 1.1).

Switching to renewables will not suffice though to get our societies off of fossil fuels. **Energy conservation** (not spending energy) **and energy efficiency** (deriving more work from the same amount of energy spent) will play an even more important role in the dearly needed energy revolution. In fact, energy conservation and efficiency might be one of the two main levers which could allow for **a decoupling of sustainable economic development from resource consumption** (see 1.2). The other major lever will be applying **circular economy** principles to our now strictly linear economy (see 3.1).

1.1 Renewable energy sources for your household (electricity and heating)

If possible, switch to an energy provider (or a tariff offered by your current provider) that supplies **electricity from 100% renewable sources** (water, wind, solar, and others).

If possible, switch to an energy provider or tariff which offers **heating from the least carbon-intensive source**. A renewable option would be **heat from biomass** power plants (mostly available in rural areas). The least carbon-intensive in urban areas is **district heating** (where available).

If heating in your area is only possible based on fossil fuel consumption, **natural gas is the least carbon-intensive** option. Heating with oil and coal, on the other hand, is highly carbon-intensive and should be phased out as quickly as possible.

If you live on self-owned property or have an opportunity to do so, consider installing **photovoltaic modules** for producing your own electricity. A **solar thermal pump** may provide you with free warm-water processing. For heating you could install an **air-air heat pump**, a **ground-source heat pump**, or a **pellet furnace**. Either of these options will not only shrink your carbon footprint but but also your energy bill. Installation costs will be redeemed quite quickly.

1.1 Transition to renewable energy sources

Renewable energy sources such as sunlight, wind, moving water, biomass, and geothermal heat have two unassailable advantages: 1. They constantly renew themselves—no human contribution is required. 2. They exist in such abundance that mankind could never fully use them.

Consider the power of the enormous nuclear fusion reactor at the center of our solar system, the Sun: It will go on fusing atoms for another five billion years. Even all of today's gigantically risen global primary energy demand of almost 160,000 terrawatt-hours per year (see p. 75) equals only ca. 0.01% of the solar energy received on Earth's surface each year.

Apart from their inexhaustibility, renewables have also become commercially viable: "Thanks to years of active policy support and driven by technology advances, rapid growth and dramatic reductions in costs of solar photovoltaics (PV) and wind, **renewable electricity is now less expensive than newly installed fossil and nuclear energy generation** in many parts of the world; in some places it is less expensive even than operating existing conventional power plants."[106] Many economists believe that with *Levelized Costs of Electricity* (LCOE) of solar and wind energy now having dropped below those of fossil fuels and nuclear it is only a question of time until the latter will become a burden also commercially.

Besides being inexhaustible and having become commercially competitive, renewable energy has another major advantage: In contrast to nation- or region-wide power grids, which depend on centrally controlled coal, gas, or nuclear power plants, renewables can be deployed in smaller units, independent of large grids. This is of particular importance to the **1.1 billion people worldwide who do not have access to electricity grids** at all (particularly in sub-Saharan Africa and parts of Asia). There, **microgrids** powered by various renewable energy sources or even single-household insular PV modules with attached storage can (and already do) provide electricity. In such regions, renewable energy in small units allows for **"leap-frogging"** over the decade-old technology of centralized power grids. In industrial and post-industrial economies ever more pervasive digitization allows for a similar coup: Many **distributed micro-producers** of electricity (with PV modules on their rooftops) can be **crosslinked into a smart grid**, thus simultaneously becoming producers, consumers, and distributors.[107]

Some societal and technical challenges remain to be tackled though before renewables can completely supplant fossil fuels in electricity production. While renewable energy sources are inexhaustible, the means to harvest and harness them are not, primarily because of topographical restrictions. There are only so many opportunities to build large-scale windparks (on- and offshore), hydropower dams, and solar arrays before such installations become environmental liabilities themselves. Resistance of local populations against ever more energy-related building projects also has to be taken seriously. Geographical restrictions also affect installations, which are not built for harnessing renewable energy per se but rather for storing it. But large-scale storage solutions for electricity are dearly needed. In fact, storage is the main technical challenge to be overcome before renewables can take over. While solar photovoltaic and wind can principally generate enough electricity to power even high-consumer societies, they cannot do it *all the time*. The Sun only shines a few hours a day, depending on weather and season, and wind may also only blow intermittently. Yet, around-the-clock availability of electricity is essential. Present-day power grids deliver exactly the amount of electricity that is needed at any given point in time during the day or night. They are able to do so because energy output is generated by conventional coal- or gas-fired as well as nuclear power plants. Their output can be ramped up and down quickly (within a matter of minutes). An electricity grid based on conventional power plants can therefore always carry *base load* (the average amount of electricity needed at any day- or nighttime) as well as *peak loads* (peak amounts are needed primarily in the mornings before people leave their homes and in the evenings when they return home). The abundant energy that renewable sources can deliver needs to be stored *when they deliver*, so that it is available *when it is needed*.

Today, primarily pumped-storage power plants are employed for energy storage. When renewable sources deliver energy to them, they pump water from reservoirs at low elevation to reservoirs at higher elevation. When energy is needed, the stored water is released and drives hydropower turbines, which generate electricity. Obviously such installations are only feasible where water reservoirs at different elevations are possible (as in mountainous regions). But even there, building as many such installations as would be needed is out of the question. The concurrent destruction of natural ecosystems is prohibitive. Existing battery technologies (like lithium-ion) are also in use for large-scale storage, but battery production has itself become an environmental issue (see p. 112) due to the damaging extraction of the non-renewable raw materials needed (primarily lithium and cobalt but also rare earths). A high recycling/remanufacture ratio for battery materials will be essential in the near future.

Other existing storage solutions include power-to-gas (P2G), which converts electrical power to methane. First, electricity is used to split water (H_2O) into hydrogen (H) and oxygen (O)—this process is called electrolysis. Then, the resulting hydrogen is combined with carbon dioxide (CO_2) to form methane (CH_4), which is stored in large tanks. All in all though, converting electricity to a fossil fuel like gas might not be the most desirable storage solution in the future anyway. Far more interesting is the hydrogen produced through electrolysis: It can be used of itself as a fuel in fuel cells but also as a storage solution (see pp. 194–197).

Concentrated solar-thermal power
COULD SUPPLANT COAL- AND GAS-FIRED AS WELL AS NUCLEAR POWER IN ELECTRICITY GRIDS.

→ **Concentrated solar-thermal power plants (CSP)** like this one in Tonapah, Nevada, US, can produce hundreds of thousands of megawatt-hours of electricity per year. They use molten salt storage to make that energy available day and night. The largest such facilities are currently located in China, Dubai, Morocco, and the US. Utilizing high voltage, direct current (HVDC) electric power transmission, the energy produced at CSP plants can be delivered losslessly over thousands of kilometers.
Photo: iStock.com / Mlenny

↓ **Heliostats** at a CSP plant
Photo: Dennis Schroeder / NREL (CC BY-NC-SA 2.0)

Wind turbine, India. This turbine is one of many of a 17.5 megawatt wind project in India, consisting of eighteen wind farm sites in the Indian states of Tamil Nadu, Maharashtra, Rajasthan, and Karnataka. The turbines supplant electricity received from the Indian National Grid, which is dominated by coal and oil-fired power generation. Photo: Land Rover Our Planet (CC BY-ND 2.0)

Fortunately, there are renewable energy technologies that do away entirely with external storage needs. In fact, they work the same way as conventional coal-fired, gas-fired, or nuclear power plants do in electricity grids. One such technology already in wide use are concentrated solar-thermal power (CSP) plants (images p. 191). CSP plants use large arrays of mirrors to concentrate sunlight on a central tower, where that light is first converted into electricity, and then into thermal energy. This thermal energy can be stored in molten salt (or another suited medium) for up to 15 hours. CSP plants are thus capable of carrying base *and* peak loads. Large-scale CSP installations exist in sunny, desert localities in Spain, the US, and Morocco. Countries in the Middle East, India, and China are also rapidly building up CSP capacity. But CSP plants do not only work in desert countries: Using high voltage, direct current (HVDC) electric power transmission, they can losslessly deliver the energy they produce over thousands of kilometers.

While all the aforementioned renewable energy production and storage technologies will mature even further in coming years and technical challenges will be overcome, they cannot drive the required energy revolution by themselves. When we are talking about renewable energy we are focussed almost exclusively on *electricity generation*. But electricity only accounts for ca. 20% of end-use energy today (primarily in buildings, some in industry). 65–70% of

1.2 Invest in your home's energy efficiency and conserve as much energy as you can

If you own or live in a building with bad **thermal insulation**, consider refurbishing it to become more energy-efficient, or petition the owner of the building to do so. There might be national or regional government grants to support such energy-saving renovation measures.

Use **energy-efficient household appliances** with an A+ or better still an A++ rating. This tip holds especially true for appliances that run all the time (like your refrigerator or air conditioner) or appliances with high energy requirements (like washing machines, microwave ovens, TV sets, and computer monitors).

Switch from your old filament or energy-saving light bulbs to far more **energy-efficient LED-lights**.

Don't leave lights on in rooms you are not in.

Don't leave electrical appliances running on **standby mode** (which still consumes energy that you have to pay for).

In winter, don't leave windows open for long to air spaces. A few minutes are enough.

For most clothes and their typical use it is more than sufficient to **wash** them at 30 or 40 °C. Instead of using a tumble-dryer hang your clothes to dry.

Take **fewer baths and shorter showers**. Water is as much a valuable resource as the energy you need to heat it up.

When **washing dishes**, don't leave the water running. You only need it while rinsing dishes.

end-use energy are delivered by fossil fuel combustion: ca. 40% by oil in the transportation sector and ca. 25% by natural gas and oil in buildings and in industry. The rest of end-use energy is delivered by nuclear, biomass, and other renewables such as biodiesel. Therefore, a true energy revolution will have to go much further than just supplanting fossil fuels in electricity generation.

1.2 Energy efficiency and conservation

While some technologies challenge fossil fuel combustion in transportation (see pp. 194–195), in heating buildings, and in industrial processes (see p. 196), renewable energy sources are still a far cry from supplanting fossil fuels in these sectors. The real game changer here will be drastic increases in **energy conservation and energy efficiency**. What is needed is not so much an energy revolution but an **efficiency revolution**. Recent estimates show that a five-fold increase in efficiency is needed *and* possible, as Ernst Ulrich von Weizsäcker and Anders Wijkman highlight in *Come On!*, their Report to the Club of Rome.[108]

The most obvious example of conserving energy and/or increasing its efficiency is **improving thermal insulation of existing buildings (retrofitting)**, which lowers energy consumption for heating and cooling. New housing, office and industrial spaces built to **low-energy or passive house standards** can dramatically reduce energy demand in buildings down to 5–10% (!). *Active houses* with their own renewable energy (like PV roofs) can even produce more energy than they consume. Beyond buildings, the foremost energy-consuming products powered by fossil fuels are **transportation vehicles**: cars, trucks, airplanes, and ships. While we will cover alternative powertrains in the mobility section (see p. 194), vehicles also definitely need a lot of reconsideration in terms of energy efficiency. For one, they need to become lighter so as to reduce deadweight, and secondly, fuel efficiency in general needs to improve dramatically: 86% of energy produced in a typical petrol or Diesel engine vehicle never reaches its tires. It is just lost inside the combustion engine.

We know energy efficiency ratings mostly from consumer products such as electrical appliances, lighting, computers, and electronic devices. While efficiency gains for these products are important, again, they just concern electricity use. But it is in the mainly fossil fuel-powered sectors (transportation and industry) where only conservation and efficiency gains will allow for a **decoupling of natural resource consumption from sustainable economic development**. Therefore, let's take a closer look at these two sectors now.

2. Mobility

2.1 Individual transportation

We already mentioned that automotive vehicles for individual transportation need to become much more fuel-efficient, regardless of what their fuel may be. Smaller and lighter ("fitter") vehicles with more carbon-fiber parts benefit petrol engine vehicles (more mileage per fuel unit) as much as they do electric vehicles (extended range).[109] Unfortunately, the trend for the past ten to fifteen years has been for larger and heavier vehicles, particularly so-called *sports utility vehicles* (SUVs). SUVs are a class of cars non-existent until recently—no one had a need for them. Automakers will need to be incentivized or forced by regulation to rethink that trend they initiated. In Europe, car manufacturers now have to do so anyway to some extent because they need to diminish their cumulative fleet CO_2 emissions as per EU regulation. 2020 is already seeing a flourishing of new models of battery electric vehicles (BEVs) or plugin hybrids (PHEVs). But new BEVs and PHEVs are only going to get automakers so far: Petrol and Diesel engine vehicles will also have to become much more fuel-efficient in coming years.

2.2 Public transportation / Freight by road, rail, and sea / Aviation

While it now seems settled that the electric drivetrain will be the go-to solution for future automotive transportation, a viable alternative to battery storage of electricity has been around for a long time. But **hydrogen fuel cells** never gained as much traction on mobility markets as did BEVs—despite their major advantages. Highly compressed hydrogen gas (H_2) from a tank is combined with ambient oxygen (O) inside the fuel cell, which then produces electricity and water (H_2O) from these ingredients on the go. The electricity powers the motor, and pure water vapor exits the vehicle's tailpipe. The fuel cell process is a win-win proposition on even more levels: Compressed hydrogen has an energy density much higher than that of a lithium-ion battery. That is why even from relatively small amounts of hydrogen you get decent ranges in **fuel cell electric vehicles (FCEVs)**. The latter advantage also makes fuel cells interesting for use in airplanes. In aviation, electricity stored in batteries will never take off commercially—in a literal sense: The ratio of increased battery weight to power and range gains would always keep heavy jet aircraft grounded.

In addition, fuelling a hydrogen gas tank only takes a few minutes, while recharging the battery in a BEV takes from at least half an hour (at a high-capacity quickcharger) to hours (on a regular charging station). So why has the fuel cell not long overtaken the cumbersome battery electric setup?

For one, producing the required highly compressed hydrogen gas (or liquid hydrogen) from water and electricity through electrolysis (the reverse process of what happens inside the fuel cell) is quite an energy-intensive, energy-inefficient,

Local. Walk, ride a bike, and use public transportation in your city whenever you can. Not only are these modes of transport going to lower your carbon footprint—everyday physical activity is also the best remedy against diseases of civilization like obesity, cardiovascular disease, postural defects, and back pain. When in need of a car in a city, consider taking a taxi, using commercial car sharing, or arrange for private car sharing. The typical European car is parked 92% of the time.

Regional. If there are no train connections and you need your own vehicle for commuting to work, see if you can use public transportation for at least part of the way (park & ride). Not only will that lower your carbon footprint but you will also contribute to less car traffic, congestion, and air pollution in the city. Try carpooling—it can easily be more fun and less stressful than sitting in your vehicle alone. When you buy or lease a new vehicle consider its carbon footprint over its entire life cycle (see p. 111).

National / international / intercontinental. Instead of flying, consider taking the train (or car train) or bus to your national or inner-continental destination. Short-haul or domestic flights (up to 3 hours) emit more than twice the amount of CO_2 per passenger kilometer travelled than do long-haul flights (6–12 hours, see p. 111). That's due to the fact that take-off and landing consume the most fuel. In general, by flying you emit about a 100 times the amount of CO_2 you would by taking the train (see p. 111). As long as flying is not properly carbon-taxed, offset your carbon emissions for unavoidable flights via a platform like gold-standard.org.

Bicycle rush hour, Copenhagen, Denmark
56% of Copenhageners ride a bicycle
every day. 75% cycle all through winter.
Photo: Mikael Colville-Andersen / Copenhagenize
Design Co. (CC BY-NC 2.0)

and thus costly process so far. This is why today most hydrogen is produced using cheap fossil fuel energy! In that sense, the carbon footprints of both BEVs and FCEVs depend a lot on how the electricity going into the battery, respectively into hydrogen production is generated (see comparison p. 111). In the short term, efficiency gains in electrolytic hydrogen production are to be expected though, and even more importantly: The energy required for electrolysis could be derived entirely from unused peak electricity in windparks. **Wind turbines could simultaneously produce electricity and hydrogen, and may thus even offer an energy storage solution** in the near future.

The other reason why hydrogen-powered FCEVs have not made a grand market entrance in most of the world is a lack of fuelling infrastructure. While fuel cell vehicles give users much better ranges than BEVs, the number of hydrogen fuelling stations is still extremely small as compared to electric charging stations. This might change though with growing policy and industry support.

While the scarcity of fuelling infrastructure may still make fuel cell vehicles impractical for everyday individual use, it does not for vehicles with high daily mileage going along predetermined routes: **heavy-duty and/or high-speed vehicles like freight trucks, public buses, airplanes, and ships can all be powered by hydrogen fuel cells.** Japan along with the world's largest automaker Toyota is intent on becoming the first hydrogen-powered nation in the world. Toyota already produces hydrogen-fuelled cars and, in collaboration with Japanese company Hino, also buses (the Sora) and heavy-duty trucks (see image p. 197).

For its FreeCO$_2$AST project, Norwegian company Havyard is currently developing the world's largest **fuel cell cruise ship**. Several companies around the world and the NASA-sponsored CHEETA project (Center for Cryogenic High-Efficiency Electrical Technologies for Aircraft) are working on **fuel cell electrical airplanes**, which might debut as early as the mid-2020s.

For all these means of transportation hydrogen could become the number one solution for completely supplanting fossil fuels in the near future. Besides Japan, fossil fuel behemoth Australia is now developing a national strategy for becoming the world's largest hydrogen producer in the coming decades; and fossil fuel dwarf Austria is also planning to apply hydrogen technology, primarily in industrial processes: its major steel producer VOEST Alpine is currently testing a CO$_2$-free hydrogen production facility (the EU-funded H2FUTURE project), whose hydrogen output could supplant fossil fuels in steel production. That would be no small feat, as VOEST is currently responsible for 10% of Austria's total CO$_2$ emissions.

Besides applying hydrogen technology in the fields of road and maritime freight, aviation and public transportation, as well as in industrial production, technological advances will have to be complemented by a complete eradication of fossil fuel subsidies in these sectors (see pp. 203–204). Money thus saved by governments needs to be spent on **extension** (reaching more places), **intensification** (introducing shorter intervals), **and price reductions in public transportation**. This applies to urban, regional, national and international train and bus connections for passenger transportation (e.g. inner-continental high-speed rail connections). Further measures are also necessary in **freight traffic**, which **has to become increasingly mutimodal** in the future, allowing for rapid and economic switching of transportation means along routes (e.g. truck to rail and waterways).[110]

Smart mobility systems based on deep digital connectivity can also help intensify **multimodal public passenger transportation**. A real-world application in the city of Vilnius, Lithuania, called "Trafi" shows what that can entail. All means of public transportation, car sharing services, weather and traffic information are connected and displayed in real-time via the mobile phone app, including one-touch payment for any mode of transportation. Users can chose a route depending on how fast they need to be, price, or other criteria. For example, for the fastest route the app might propose you take bus #43, which will arrive 75 m from your current location in 2:30 minutes, get off the bus at the third stop, where an electric sharing vehicle is parked, which you can book and pay for on the go, drive that car to within 300 m of your destination along the currently least congested route, park the car there, and walk the rest of the way as there will be no public parking at the destination. That sounds smart indeed, and the system works so well that it is already being exported to other cities in Europe.

→ Wind turbines

PRODUCE ELECTRICITY.
BUT WHEN THEIR PEAK
POWER IS NOT CONSU-
MED, THEY CAN ALSO
PRODUCE HYDROGEN
THROUGH ELECTROLYSIS.

Installation of a wind turbine
in Upper Thumb, Michigan, USA.
Photo: Consumers Energy (CC BY-NC-SA 2.0)

← Hydrogen

CAN POWER ALL
KINDS OF FUEL
CELL VEHICLES.
BUT IT CAN ALSO BE
USED AS A MEANS
OF ENERGY STORAGE.

Heavy-duty fuel cell truck,
running on hydrogen, developed by
Japanese automakers Toyota and Hino.
Photo: Courtesy of Toyota Motor Corporation

3. Industrial production:
Linear vs. circular economy

The globalized world we live in today runs on a **linear industrial economy** (also called a **throughput economy**): Natural resources constantly flow in on one side of the system as new material, get processed and manufactured into products and goods, which are then marketed, sold, consumed, and ultimately discarded as waste on the other side. This is the linear "take—make—waste" approach. "From 1970 to 2017, the annual global extraction of materials [from Earth] tripled and it continues to grow, posing a major global risk."[111] Although some used goods get recycled, and the embodied materials reenter the resource flow, in classical recycling they only do so at drastically reduced rates and highly depreciated (i.e. at much lower quality and value). In one word: materials are only *downcycled*. In European industry "[o]nly 12% of the materials it uses come from recycling."[112] In other words: In a linear system **sustained economic growth** can only be achieved by also keeping material input and throughput at that same rate of growth.

Deceivingly, this model seemed sustainable during the industrial era and for decades of greatly accelerated post-WWII economic development: both on the input side of the linear model, where seemingly infinite natural resources flowed in, as well as on the output side, where waste from production and consumption was simply dumped on nature as "externalities." Now, both in regard to resource inputs as well as to polluting outputs, we have long entered a **phase of overstepping planetary boundaries (see pp. 150–155).** Neither can finite Earth give as many resources as would be needed to sustain linear economic growth, nor can she take more pollution, waste, and accompanying destruction of natural ecosystems before she will stop functioning the way we expect her to.

Just as with energy, which only conservation and drastically increased efficiency can keep available at the required levels (see 1.2), natural resources and materials as well can only be kept available at current high levels if they are conserved and used more efficiently. There is really only one way to achieve this: The linear industrial economy has to give way to a **circular economy (CE)**, in which resources, materials, and products are kept in use cycles at constant high values (i.e. without getting depreciated in downcycling spirals). "A circular economy is one that is restorative by design, and which aims to keep products, components and materials at their highest utility and value, at all times."[113]

CE differentiates materials as being either *biological* (meaning they can be safely decomposed in the biosphere) or *technical materials* (non-biodegradable, kept at high quality and away from the biosphere in their own industrial cycle). **In CE no or hardly any waste accrues.** In the **biological cycle**, biological materials that have not been consumed or biological "waste" become "food" for uses further down the line: Energy may be recovered from them (as in the case of

3. Make lasting use of the things you own

Keep goods and products for as long as they work. If a product does not function properly anymore, see if you can repair it yourself, or if you can have it **repaired** or **refurbished**.

Buy higher-quality goods and products which will last longer and come with **extendable or even life-long warranties**. Buy products which are designed in a way so that they can easily be upgraded or refurbished later on in their life cycle.

Recycle as many of the goods and products you no longer want to keep as possible instead of discarding them as waste. Besides public waste recovery facilities, some manufacturers also offer **product recycling programs**, where you might redeem some of the value you originally paid for. Consumer electronics are sometimes refurbished by manufacturers and sold again in secondary markets.

Consider buying **products designed to be refurbishable**, like the decidedly modular Fairphone: Not only is it produced sustainably, using conflict-free and fairly traded minerals as well as recycled copper and plastic (see p. 134). It is also highly repairable with most of its components easily changed, exchanged, or upgraded later in its product life cycle. Fairphone ranks #1 in Greenpeace's Consumer Electronics guide (see GREENPEACE [2017]), with only Apple's iPhone as a close #2 runner-up. Apple has intensified its takeback and recycling program in recent years, employing recycling robots to disassemble and recycle the materials used in their phones.

Consider not following the latest fashion and not buying **clothes** you might not wear often or perhaps only once. Pay attention to where clothes have been manufactured and under what conditions. If the manufacturer or reseller do not provide such information, consider not buying their products. When ordering cheap clothes via online retailers, which offer free takeback, be aware that the clothes you send back might not be "put back on a shelf" at all but might simply be discarded as waste. In general, "[i]t is estimated that more than half of fast fashion produced is disposed of in under a year." See ELLEN MacARTHUR FOUNDATION (2017), p. 19.

THE TWO USE CYCLES OF THE CIRCULAR ECONOMY

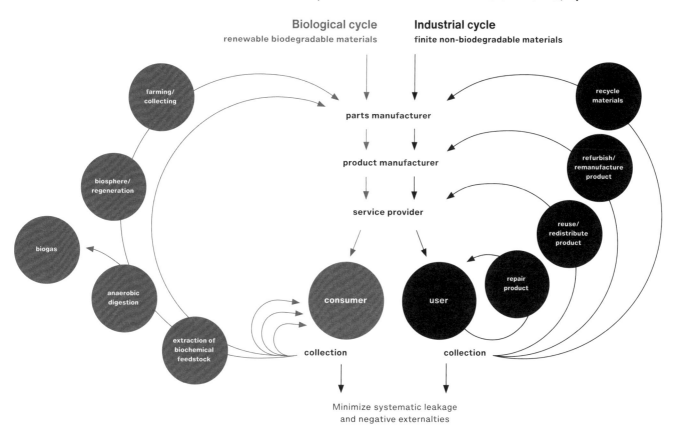

Biological cycle
renewable biodegradable materials

Industrial cycle
finite non-biodegradable materials

farming/collecting

recycle materials

parts manufacturer

biosphere/regeneration

refurbish/remanufacture product

product manufacturer

service provider

reuse/redistribute product

biogas

anaerobic digestion

consumer

user

repair product

extraction of biochemical feedstock

collection

collection

Minimize systematic leakage
and negative externalties

Graphic: SCHIENERL D/AD,
based on WEBSTER (2017), p. 19

biomass power plants), or biochemical ingredients may be extracted and turned into biogas or nutrients for restorative natural processes, which then produce new biological materials (see above) In the **industrial cycle**, on the other hand, parts and products are maintained in use for as long as possible, get repaired if they don't work anymore, get reused/redistributed in secondary or tertiary markets, get refurbished/remanufactured to become new parts/products, or ultimately get recycled into new materials if parts/products have lost all utility. Of course, industrial products and parts need to be designed for easy disassembly, repairability, refurbishment, and upgradeability right from the start. Growth- and profit-maximizing cheats by linear economy actors (like ever shorter product life cycles through *planned obsolescence* and near-impossible repairability of products) are debunked as plain corporate delinquency when compared to CE practices. CE principles are condensed in the "Cradle to Cradle" concept (and the book of the same title) by American architect William McDonough and German chemist Michael Braungart.[114] You can find more information on the

Cradle-to-Cradle (C2C) concept in the short section about how this book was printed (C2C-certified[115]) on page 215.

The circular model lends itself particularly well to establishing a **performance economy** in place of today's production/consumption economy. Rather than being consumers who buy and own products until and beyond the end of their life cycle (at which point consumers assume responsibility for the waste these products now constitute), customers in a performance economy rent or lease the *utility* or *performance* of products. Renting cars for mobility is an obvious example. But in a true performance economy, customers might also lease furniture, their washing machine, building materials like steel beams, or even carpets. The company leasing products to customers maintains ownership of them and remains responsible/liable for their functioning. Once the customer does not need a product anymore, the company takes it back, refurbishes it, or disassembles it into parts, or fibers, and makes new products from those.

Nature has always only known the circular economy model: Everything gets reused, no such thing as waste exists in natural processes. But CE is not a new human idea either. „It has been with mankind for as long as we have existed", says Walter Stahel, Swiss economist and one of the main proponents of circular economy concepts. Particularly in times of resource scarcity, circular concepts have been applied to human economy time and again.

The Ellen MacArthur Foundation investigates a variety of economic fields, for which circularity could (and already does) work, and collaborates with major global corporations from these fields to realize circularity in their operations.[116] In the European Comission's Green Deal, circular economy features prominently as one of the main ways out of the climate crisis. The Green Deal aims to achieve no less than a decoupling of economic growth from resource use,[117] and it promises a circular economy action plan to achieve this.[118]

4. Food / agriculture / land-use

There is a strong connection between *how* certain foods are produced nowadays (industrial agriculture), *what kind* of food is produced (meat and dairy), and how Earth's arable *land is (over-)used* (land-use change) *and degraded* (soil degradation due to excessive use of chemical fertilizers, herbicides, pesticides). "Food production . . . results in air, water and soil pollution, contributes to the loss of biodiversity and climate change, and consumes excessive amounts of natural resources, while an important part of food is wasted. At the same time, low quality diets contribute to obesity and diseases such as cancer."[119]

To keep things simple let's just consider these two numbers: **Nearly 80% of global farmland are used for raising livestock (as grazing land and for feed**

4. Food / grocery shopping

As much of the food you and your family needs should be **locally sourced** (up to 300 km) **and organically farmed, preferably seasonal**, and **the least processed and least packaged** you can find. Buying at farmers' markets or from farming cooperatives, which often offer online ordering these days, will almost certainly let you tick all of the above boxes.

Buy only as much food as you actually need and eat. **Food waste** and food losses are a real problem, occuring along the whole chain from harvest, processing, and wholesale to retail and household consumption.

Eat as **little animal-sourced food** (meat and dairy products) as possible. Poultry and fish are the least carbon-intensive meat products. The production of red meat from pigs and ruminants (cows, sheep) requires an enormous amount of resources (land, feed, water, energy). So do the dairy products these animals produce. Once you conceive of these foods as special treats and valuable, you might also be willing to pay more for organically farmed products of higher quality. After all, the food you eat will become part of you (if only temporarily).

"A **planetary health plate** [see *planetary healthy diet* in the main text] should consist by volume of approximately half a plate of vegetables and fruits; the other half . . . should consist of primarily whole grains, plant protein sources, unsaturated plant oils, and (optionally) modest amounts of animal sources of protein." See EAT-*LANCET* SUMMARY (2019), p. 9.

For a lot of meat and dairy products **plant-based alternatives** exist today: meat-imitating vegetal burger patties or sausages have mostly become indistinguishable from their meat-based counterparts in terms of looks and taste. Cow milk can easily be supplanted by protein-rich plant drinks like soy or oat milk.

Avoid using **one-way plastic packaging and bottles**. Opt for returnable and reusable glass bottles and containers when available. Bring your own containers to the supermarket for whatever unpackaged food you may be offered there.

Recycle as much of the unavoidable household trash you generate as possible. Residual waste will either end up in a landfill or ideally be incinerated for thermal recovery. But the best trash is the one you simply avoid while shopping.

production) but only 18% of the global calorie supply consist of meat and dairy foods.[120] Large areas in the Amazon, the Congo Basin, and in Indonesia (the three largest rainforest areas on Earth) have undergone rapid deforestation in recent decades, which leads to increased GHG emissions from land-use change (LUC). Monocultural crops like soy and wheat (as well as oil palms) are cultivated on these newly deforested lands, destined to become globally distributed feed for industrially raised cattle, pigs, sheep, and chickens. 90% of soy production worldwide serve exclusively that purpose. Human consumption of soy in the form of tofu or soy milk makes up just a small fraction. While these forms of industrial agriculture use up enormous amounts of natural resources, they play a minor role in feeding the world: "Family farming is the predominant form of food and agricultural production in both developed and developing countries, producing over 80 percent of the world's food in [nutritional or caloric] value terms."[121] Feeding a projected world population of 10 billion people in 2050 (currently 7.8 billion) will be an enormous challenge whichever way you look at it. But given that current unsustainable practices in agriculture and the land-use change it requires are responsible for 25% of anthropogenic greenhouse gas emissions, require 70% of global freshwater, and are the main driver of biodiversity loss, feeding that many people in such a way would lead to a collapse of the biosphere.

Based on the best available science, the 2019 "EAT-*Lancet* Commission" report, lead-authored by Johan Rockström and Walter Willett, is the most comprehensive study to date of how to achieve healthy nutrition for all mankind from sustainable food systems in the future.[122] To achieve this "[t]he readily implementable actions investigated by the Commission were: 1) A global shift toward healthy diets; 2) improved food production practices; and 3) reduced food loss and waste."[123] In this context, *healthy diets* means "increased consumption of plant-based foods—including fruits, vegetables, nuts, seeds and whole grains—while in many settings substantially limiting animal source foods."[124] A call for **peak livestock** (the point in time from which worldwide livestock production would not continue to increase) within the next few years is sounded ever more loudly and clearly these days: "Reducing meat and dairy, and eating plant-based diets instead, would free up land to be returned to natural forest."[125] Restoring natural vegetation, such as forest, is currently also the best option at scale for removing CO_2 from the atmosphere (**carbon capture and storage, CCS**).[126] Potential areas for **reforestation** (forestation of formerly forested areas) and **afforestation** (forestation of hitherto unforested lands) are estimated at 0.9 billion hectares worldwide, which could store 205 billion tonnes of carbon (equalling 752 billion tonnes of carbon dioxide).[127]

But healthy diets alone will not suffice to guarantee supply of enough sustainably produced food for 10 billion people in 2050. According to a 2011 FAO study,[128] **food losses and food waste** occuring along the chain from agricultural

production, food processing, wholesale, and retail to final consumption amount to **one third of food being destroyed and lost**. From malpractices in harvest and insufficient transport and storage facilities, overproduction and excessive standardization of food appearance to bad food planning in households or over-cautious minding of expiration dates food losses and waste mount.

Improvements in food production practices would mainly result from chemical-free intensification of crop yields. This could entail applying natural principles of soil nutrient regeneration as it is done in *permaculture* or in *carbon farming* and redistributing remaining stocks of natural mineral fertilizers to world regions that suffer from less or no soil degradation (e.g. large parts of Africa).

If principles like the ones delineated here were to be applied, a firm link between human and planetary health would be established via a **planetary health diet**: Eating less meat, dairy, and sugar- and palm-oil-containing food would mean that diseases such as obesity, high cholesterol, diabetes, and consequently cardiovascular diseases could be pushed back, while health would also be restored to the planet and its biosphere, which would suffer less biodiversity loss and could sequester more greenhouse gases on reforested/afforested lands.

Just as is the case with fossil fuel use, a substantial shift in subsidizing policies is also necessary regarding agricultural production: away from large-scale industrial agriculture including factory farming of animals to small-scale regional and organic food production. The role of policies and regulation in all economic and ecological sectors will be highlighted in the next section.

5. Public sphere / policies / regulation

While we don't have any direct influence on policies and regulation as individuals, we can certainly **make use of our right to vote** and back parties and candidates who have environmental issues on their agenda and boast a track record in that regard.

There are other ways for an individual to make an appearance in the public sphere. Greta Thunberg has powerfully demonstrated that "no one is too small to make a difference." What began as a solitary **school strike** in front of the Swedish parliament in August 2018 turned into a global mass phenomenon within a matter of months. In 2019, three global Earth Strikes brought hundreds of millions of young and older people to the streets all over the world (see p. 206). Maybe you are already part of that movement—if not, join your local group of Fridays for Future, the Sunrise Movement, Extinction Rebellion, or another such alliance. Take to the streets together, raise your voices, and **make your protest heard and seen**. That kind of visibility and political pressure from the streets is at least as important as the power we collectively hold as consumers.

Sign petitions to your local government or even to world leaders via avaaz.org.

Join class-action lawsuits against companies or governments that are oblivious to their liability toward the commons and society at large. I recently joined such a class action, instigated and organized by Greenpeace, which demands equal treatment of airlines and rail operators in my country. In Austria, air tickets are exempt from value-added tax, whereas rail tickets are not. What argument could possibly justify such discriminatory treatment? Higher-stakes lawsuits are brought against globally operating corporations by sub-national, national, or supranational governing or legal bodies more often these days: think of the case against ExxonMobil brought to court by the New York and New Jersey Attorneys General (see p. 160–161) or the class action suits brought against Volkswagen in the wake of the emissions scandal.

Generally, the classical lines of conflict seem to become more blurry in combatting something as all-pervasive as the climate crisis: US federal states now openly oppose and refuse to comply with environmentally harmful measures taken by the federal government under Trump. Greta Thunberg was invited as a speaker to the World Economic Forum in 2019 and in 2020 in Davos,

The climate crisis is a global one. As such it cannot be tackled by single economic, political, or societal actors but only by the broadest possible coalition of forces all around the globe. Nonetheless, nation states and supranational governing bodies like the European Union have a special responsibility in creating the necessary regulative and policy frameworks that allow for an effective tackling of the crisis. So do international political and economic organizations like the United Nations or the OECD. Multilateral treaties like the Paris Agreement support such a framework. Delineating a roadmap for its Green Deal, the European Commission puts it as follows:

"The environmental ambition of the Green Deal will not be achieved by Europe acting alone. The drivers of climate change and biodiversity loss are global and are not limited by national borders. The EU can use its influence, expertise and financial resources to mobilise its neighbours and partners to join it on a sustainable path. The EU will continue to lead international efforts and wants to build alliances with the like-minded."[129]

Economic "institutions" like free markets and the financial sector also hold enormous weight and power in achieving the required transition. Yet they cannot be expected to "do the right thing" of their own free volition. Free market and financial actors are not accountable to the public or the greater good but mostly to shareholders of privately owned companies and corporations. Their primary objective is to maximize profits and profitability of these economic entities. Considering that neoliberal free-market policies and underregulation of the financial sector during the last four decades coincided with unprecedented levels of greenhouse gas emissions and a global financial crisis, free market forces and financial actors have to be "incentivized" and regulated by political institutions so that they will contribute a fair share of their enormous power to effectively addressing the climate emergency.

National and international policies and regulative measures adequate to that end, including taxation, subsidies, legal restrictions, and bans, are too numerous to be covered here, but the following list contains the most discussed and supposedly most effective measures today:

• **Phasing-out of fossil fuel subsidies**. "Historically, subsidies granted to the fossil fuel industry were designed to lower the cost of fossil fuel production and incentivize new domestic energy sources . . . At a time when renewable energy technology is increasingly cost-competitive with fossil power generation, and a coordinated strategy must be developed to mitigate climate change, the broader utility of fossil fuel subsidies is being questioned."[130]

Depending on whether a narrow or a broad definition of the term *fossil fuel subsidies* (FFS) is applied, global FFS amount to from between several hundred billion US dollars per year (2017: 340 billion USD)[131] to a mind-boggling 5.3 trillion USD per year, if all kinds of costs associated with fossil fuel use are considered (i.e. greenhouse gas emissions and other pollution resulting from the extraction and burning of fossil fuels). "These negative externalities have adverse environmental, climate, and public health impacts . . . Several international institutions, including the G20, the International Energy Agency, and the Organization of Economic Cooperation and Development (OECD), have called for the phase-out of fossil fuel subsidies. The European Union has also called for such a phase-out but has not yet taken concrete actions."[132] To put these financial values in perspective: The EU estimates that the transition to a climate-neutral European continent by 2050 will cost 260 billion Euros each year.[133] It seems fair then to suppose that such monies could be raised merely by phasing out fossil fuel subsidies.

"Inefficient fossil fuel subsidies can hinder progress in a country's transition [toward a lower-emissions energy system] as they distort prices, induce economic inefficiencies and poor environmental outcomes, and put pressure on scarce public resources. They can encourage the use and production of fossil fuels and the accumulation of carbon intensive assets."[134]

• **Carbon taxation**, including taxation of aviation (kerosene) and maritime fuels. Eventually, GHG emissions from all societal and economic sectors should be uniformly priced/taxed. Instead of remaining economic externalities (incidental costs of economic actions, which no one pays for), GHG emissions should be "internalized" into costs and reflected in prices. As of now, such carbon pricing exists in individual countries as well as in regions like the EU or California in the form of **emissions trading systems (ETS)**. But these systems do not yet include all sectors: The EU ETS covers only the energy-producing sector, a small number of energy-intensive industries like steel production, and inner-European aviation but not so all other transportation (road, rail, waterway), other industries, and the building sector. Too many free emission allowances and too low a carbon price (24 Euros per tonne of CO_2 emissions at the beginning of 2020) are further hampering the EU ETS so far. While becoming all-inclusive in terms of sectors, ultimately an emissions trading system should become global. During COP25 in Madrid (2019) the **establishment of rules for a future global ETS** was tried but failed due to resistance from a number of powerful national economies like the US, China, India, and Brazil. Many experts believe though that an effective global ETS could lead to rapid market-driven decarbonisation of economies, and thus become a silver bullet for tackling the climate crisis.[135]

• **Carbon customs tariffs**. While establishment of a global emissions trading system might seem far off at the moment, regional or continental "coalitions of the willing" (like the EU) could go ahead with expanding and deepening their emissions reduction and trading programs. To protect their territorial economies against

Switzerland. The CEO of the world's largest investment management firm cautions companies in their portfolio to take more serious environmental and climate protection measures in their operations—else investments might be withdrawn. While some of this may be nothing more than *greenwashing*, it nevertheless sends out signals that make it harder for climate change deniers to continue their professional filibustering and for profiteers of the fossil fuel complex to justify their exploitation and degradation of the environment (if they care to do so at all).

At the very least, **a broad coalition of obvious but also unlikely partners has come to the fore** in the fight against the climate crisis ever more visibly in recent years. This is a hopeful situation, I believe. Not everybody needs to (or will ever) be on board, when the big ship makes its big turn—a critical mass of people, NGOs, political institutions, states, and economic players is enough. Let's just do everything we can NOW so that we, this critical mass, can alter course before the pull of the vortex becomes too strong, and all we can do from that point on is to just stand by as spectators and watch the forces of nature play out their own never-ending story.

Vienna, October 2020

unfairly underpriced goods and products from countries where no comparable pricing regime exists, such coalitions could opt to impose carbon customs tariffs on imports from abroad (the European Commission calls this a "carbon border adjustment mechanism"[136]). As the world's largest single market with the largest ETS in place, the EU holds considerable global power and could set standards that apply across global value chains.[137] Growing into larger world regions or spanning multiple such regions (with China, for example), an intercontinental "coalition of the willing" might eventually form a "climate club." Such an alliance could exert considerable influence over other national economies, effectively forcing them to establish equal carbon pricing, if they want to continue engaging in trade with the climate club or become club members themselves.

- **Carbon audits.** Within states or supranational governing bodies mandatory carbon audits (also called *sustainability proofing)* should become all-pervasive. All new legislation, subsidy and financing decisions, approval procedures, and executive measures should be audited for their effectiveness in reducing greenhouse gas emissions and fighting the climate crisis.
- **Extension of existing and phasing-in of new climate-protecting subsidies.** Subsidies should be expanded for renewable energy installations and mobility, green building and retrofitting (weatherizing) of the existing building stock, adoption of circular economy principles in industrial production, small-scale regional and organic agriculture, and green investments/divestment from fossil fuels.[138]
- **Increased public spending** on the energy revolution (transition to renewable energy sources, increased energy efficiency) in the public sector and in government procurement, on public transportation infrastructure, and on reductions in pricing of public transportation would send the right signals to free markets.
- **Phasing-out of subsidies for industrial agriculture** including those for livestock production (meat and dairy) and transfer of funds to subsidizing regional organic food production.
- **Denial of new territorial extraction rights** to the fossil fuel and other extractive industries and shifting of the tax burden to polluting economic activity (and away from labor).
- **Extension of environmental protection zones** like the EU Natura 2000 zones.
- **Stricter pollution bans** (and enforcement thereof) on soil, air, groundwater, and sea-polluting chemicals and modes of production, including a total ban on one-way plastic packaging.

Earth Strike
Sept. 27, 2019
Vienna, Austria
Photo: Christian Schienerl

Notes

1. From Robert A. Heinlein's novel *Time Enough For Love: The Lives of Lazarus Long*, New York City: Berkeley Books, 1973.
2. SCHELLNHUBER (2015), 23 (in German).
3. IPCC (2018), 6.
4. See https://climate.nasa.gov/vital-signs/global-temperature/ as well as https://data.giss.nasa.gov/gistemp/graphs/graph_data/Global_Mean_Estimates_based_on_Land_and_Ocean_Data/graph.txt.
5. STEFFEN *et al.* (2018), 8256.
6. IPCC (2014), 48, fig. 1.9; STIPS *et al.* (2016), 1.
7. STIPS *et al.* (2016), 1.
8. Ibid., 1–2.
9. POWELL (2015); COOK *et al.* (2013). In their widely perceived study COOK *et al.* reaffirmed (and further proved) findings from earlier studies concerning the scientific consensus on *anthropogenic global warming* (AGW) in the peer-reviewed literature to which an overwhelming majority of 97.2% of climate scientists subscribe. As was to be expected, the study by COOK *et al.* has been criticized and questioned many times, predominantly on the grounds of denialist political and economic interests. But the data and conclusions from that study are 1. entirely transparent and 2. leave no room for doubt.
10. THUNBERG (2019), 8–9.
11. HAZEN (2012), 25.
12. LE QUÉRÉ *et al.* (2018), 2143.
13. To get an idea of how intricately complex that relation actually is, see: TOLLEFSON (2014), 276–278. Basically, a lot of atmospheric heat content is absorbed by the oceans (upwards of 90% during the past few decades). But the rate of absorption does not necessarily follow a linear path, as the Pacific Decadal Oscillation (PDO) gives proof of.
14. To learn how past temperatures can be determined accurately from the isotopic composition of water molecules in ice cores, see MULVANEY (2004).
15. IPCC (2014), 4.
16. RAHMSTORF/SCHELLNHUBER (2018), 28 (in German).
17. STEFFEN/CRUTZEN/McNEILL (2007), 615.
18. See LOVELOCK/MARGULIS (1974).
19. See Robert Hazen's excellent *Symphony in C* for the central role the element carbon plays in almost every aspect of our lives (HAZEN [2019]).
20. LE QUÉRÉ *et al.* (2018), 2143.
21. See FRIEDLINGSTEIN *et al.* (2019), 1811.
22. ROGELJ *et al.* (2019), 340.
23. WACKERNAGEL/BEYERS (2019), 252.
24. Ibid., 57.
25. Ibid., 69.
26. To learn how Footprint accounting for countries works and find out about specific countries' current Footprints, go to: data.footprintnetwork.org/; also see WACKERNAGEL/BEYERS (2019), 71–92.
27. STEFFEN/CRUTZEN/McNEILL (2007), 616.
28. RAHMSTORF/SCHELLNHUBER (2018), 16 (in German).
29. See *BP Statistical Review of World Energy 2019* (BP [2019]), 9: Renewable power sources like solar and wind accounted for only 4%, hydroelectricity for 6.8%, and nuclear energy for 4.4% of the world's primary energy consumption in 2018. *IEA Key World Energy Statistics 2019* (IEA [2019]), 6, shows 81.3% of the world's total primary energy supply for 2017 coming from coal, oil, and natural gas.
30. At current consumption rates, the proved global reserves of oil and natural gas will be depleted by the end of the 2060s (coal: 2150s). See *BP Statistical Review of World Energy 2018* (BP [2018]), 13: global proved oil reserves were at 1,696.6 billion barrels in 2017, which would be sufficient to meet 50.2 years of global production at 2017 levels; 27: global proved gas reserves were at 193.5 trillion cubic metres (tcm); this is sufficient to meet 52.6 years of global production at 2017 levels; 37: world proved coal reserves were sufficient to meet 134 years of global production, being much higher than the reserves/production ratio for oil and gas.
31. See data for 2018 in BP (2019), 55–56.
32. BAJŽELJ/ALLWOOD/CULLEN (2013).
33. EAT-*LANCET* SUMMARY (2019), 7, and WILLETT/ROCKSTRÖM *et al.* (2019), 451–452.
34. UN (2019), 1.
35. TOWNSEND/HOWARTH (2010), 32.
36. STEFFEN/CRUTZEN/McNEILL (2007), 616.
37. WILLETT/ROCKSTRÖM *et al.* (2019), 449.
38. Ibid., 449. For the most comprehensive assessment of contemporary biodiversity loss to date, see IPBES (2019).
39. IPBES (2019), 28.
40. FRIEDLINGSTEIN *et al.* (2019), 1813; LE QUÉRÉ *et al.* (2018), 2170; IPCC (2014), 3, fig. SPM.1 (d).
41. FAO/IFAD (2019), 8.
42. SMIL (2011), 619.
43. See corresponding statistics by the Food and Agriculture Organization of the United Nations visualized here: https://ourworldindata.org/land-use#how-the-world-s-land-is-used-total-area-sizes-by-type-of-use-cover; for the ratio

of agricultural land in total to land used for feed production in the US, see RIFKIN (2019), 95–96.
44. WEF (2019/1), 11.
45. IPBES (2019), 28.
46. FOTHERGILL/SCHOLEY (2019), 243.
47. WILLETT/ROCKSTRÖM (2019), 449; IPCC (2019/2), 12.
48. FAO (2011), V.
49. WILLETT/ROCKSTRÖM *et al.* (2019), 447–492.
50. See https://www.government.se/government-policy/taxes-and-tariffs/swedens-carbon-tax/, retrieved May 15, 2020.
51. As is the case with most other offsetting schemes, the carbon price that goldstandard.org is asking is only ca. 10% of what we learned would be adequate (the Swedish carbon price of 110 Euros). So I simply overcompensate by a factor of ten.
52. STEFFEN/CRUTZEN/McNEILL (2007), 614.
53. ARRHENIUS, Svante. "On the influence of carbonic acid in the air upon the temperature of the ground," *The London, Edinburgh and Dublin Philosophical Magazine and Journal of Science 5*, 1896: 237–276.
54. The expression refers to the title of Johan Rockström and Mathias Klum's book *Big World, Small Planet* (ROCKSTRÖM/KLUM [2015]).
55. See STEFFEN/BROADGATE/DEUTSCH/GAFFNEY/LUDWIG (2015).
56. STIGLITZ (2019), 14–31.
57. For this opposition between the international free-trade regime and the climate protection framework, both established in the first half of the 1990s, see Naomi Klein's excellent analysis in KLEIN (2015), 64–95.
58. CAPRA/LUISI (2016), 67.
59. For much of the following see Robert Hazen's comprehensive *The Story of Earth* (HAZEN [2012]).
60. See RICARDO/SZOSTAK (2009).
61. HAZEN (2012), 200–205.
62. For microbial lifeforms being able to survive in even the most life-forbidding conditions and environments (so-called *extremophiles*) see WARD/BROWNLEE (2000), 3–6.
63. See LOVELOCK/MARGULIS (1974).
64. For this and the following see ROCKSTRÖM/KLUM (2015), 62.
65. The *Big Five* extinction events (EE) were: Ordovician–Silurian EEs (450–440 million years ago [Ma]), Late Devonian extinction (375–360 Ma), Permian–Triassic EE (252 Ma), Triassic–Jurassic EE (201.3 Ma), Cretaceous–Paleogene EE (66Ma).
66. IPBES (2019), 12.
67. WWF (2018), 7.

68 VAN KLINK *et al.* (2020), 419.

69 IPBES (2019), 11–12.

70 Ibid., 10.

71 See *What is resilience? > Humans and nature are strongly coupled* on the website of the Stockholm Resilience Centre, retrieved Nov. 3, 2019, from: https://www.stockholm-resilience.org/research/research-news/2015-02-19-what-is-resilience.html.

72 Bats' heightened body temperature during flight and their strong immune systems protect them against falling sick with the many viruses they carry.

73 See ROCKSTRÖM *et al.* (2009). The interdisciplinary team included some of the world's leading figures in their respective sciences: Paul Crutzen, James Hansen, Katherine Richardson, Johan Rockström, Hans Joachim Schellnhuber, Will Steffen—to name just a few.

74 STEFFEN *et al.* (2015).

75 See note 53.

76 REVELLE, Roger, SUESS, Hans E. "Carbon Dioxide Exchange between Atmosphere and Ocean and the Question of an Increase of Atmospheric CO_2 During the Past Decades," *Tellus*, vol. 9, 1957: 18–27. See https://history.aip.org/climate/Revelle.htm.

77 See MEADOWS / MEADOWS / RANDERS / BEHRENS (1972).

78 SCHELLNHUBER (2015), 25 (in German).

79 Cited from a 1980 report titled "Review of Environmental Protection Activities for 1978–1979" (1–2), produced by Imperial Oil, a Canadian subsidiary of Exxon. Retrieved April 8, 2020, from: https://www.desmog-blog.com/sites/beta.desmogblog.com/files/DeSmogBlog-Imperial%20Oil%20Archives-Review%20Environmental%20Activities-1980.pdf.
A distribution list included with the report shows that it was disseminated to managers across Exxon's international corporate offices, including those in the US and the UK.

80 See "New York Sues Exxon Mobil, Saying It Deceived Shareholders on Climate Change (by John Schwartz)," *New York Times* online, retrieved Jan. 24, 2020, from: https://www.nytimes.com/2018/10/24/climate/exxon-lawsuit-climate-change.html

81 For all three warnings see https://www.scientistswarning.org/warnings/. The 1992 document was retrieved July 27, 2020, from: https://www.ucsusa.org/sites/default/files/attach/2017/11/World%20Scientists%27%20Warning%20to%20Humanity%201992.pdf.

82 UNFCCC (2015).

83 STEFFEN *et al.* (2018), 8253–8255.

84 ROCKSTRÖM/KLUM (2015), 54.

85 HOEGH-GULDBERG *et al.* (2019), 1.

86 Ibid.

87 ROMM *(*2018), 14.

88 See COUMOU/RAHMSTORF (2012), 4.

89 Calculate your own Ecological Footprint at https://www.footprintcalculator.org/, and then see "Why can't I get my Footprint score within the means of one planet?" on your results page.

90 For the following see ROCKSTRÖM *et al.* (2017).

91 UNEP (2019), IV.

92 The G20 member states are: the G7 (Canada, France, Germany, Italy, Japan, UK, USA), the BRICS countries (**B**razil, **R**ussia, **I**ndia, **C**hina, **S**outh Africa), Argentina, Australia, Indonesia, Mexico, Saudi Arabia, South Korea, and Turkey.

93 UNEP (2019), IV.

94 EUROPEAN COMMISSION (2019), 20.

95 See Climate Action Tracker's up-to-date information on countries' emission reductions (and pledges) and heating projections generated from that information: https://climate-actiontracker.org/global/cat-thermometer/, retrieved Jan. 4, 2020.

96 SCHELLNHUBER *et al.* (2012), xiii–xviii; STEFFEN *et al.* (2018), 8256.

97 climateactiontracker.org, see note 95.

98 UNEP (2019), VI.

99 STEFFEN/BROADGATE/DEUTSCH/GAFFNEY/LUDWIG (2015), 91.

100 UNEP (2019), IV.

101 See EUROPEAN COMMISSION (2019).

102 See EUROPEAN COMMISSION (2020).

103 RIFKIN (2019), 3–4.

104 Ibid., 7–8.

105 EUROPEAN COMMISSION (2019), 2.

106 REN21 HIGHLIGHTS 2018 (2019), 6; also see RIFKIN (2019), 7.

107 RIFKIN (2019), 58.

108 VON WEIZSÄCKER/WIJKMAN (2018), 144–148.

109 LOVINS (2011), 14–75.

110 EUROPEAN COMMISSION (2019), 10.

111 Ibid., 7.

112 Ibid.

113 WEBSTER (2017), 17.

114 McDONOUGH/BRAUNGART (2002).

115 See www.c2ccertified.org.

116 See https://www.ellenmacarthurfoundation.org/our-work/activities/ce100/members.

117 EUROPEAN COMMISSION (2019), 2.

118 Ibid., 7–8.

119 Ibid., 11.

120 See https://ourworldindata.org/agricultural-land-by-global-diets; FAO/IFAD/UNICEF/WFP/WHO (2020), 105.

121 FAO/IFAD (2019), 8.

122 WILLETT/ROCKSTRÖM *et al.* (2019).

123 EAT-*LANCET* SUMMARY (2019), 21.

124 Ibid.

125 See "Reach 'peak meat' by 2030 to tackle climate crisis, say scientists," *The Guardian* online, retrieved May 24, 2020, from: https://www.theguardian.com/environment/2019/dec/12/peak-meat-climate-crisis-livestock-meat-dairy.

126 See "Scientists call for renewed Paris pledges to transform agriculture (by HARWATT, Helen et al.)," *The Lancet* online, retrieved May 24, 2020, from: https://www.thelancet.com/journals/lanplh/article/PIIS2542-5196(19)30245-1/fulltext.

127 See BASTIN *et al.* (2019), 1; CROWTHER *et al.* (2015).

128 FAO (2011).

129 EUROPEAN COMMISSION (2019), 2.

130 EESI (2019), 1.

131 OECD/IEA (2019), 3.

132 EESI (2019), 1–2.

133 EUROPEAN COMMISSION (2019), 15.

134 OECD/IEA (2019), 6.

135 See RIFKIN (2019), 6–7.

136 EUROPEAN COMMISSION (2019), 5.

137 Ibid., 21.

138 Ibid., 15.

Literature

BAJŽELJ / ALLWOOD / CULLEN (2013)
BAJŽELJ, Bojana, ALLWOOD, Julian M., CULLEN, Jonathan M. "Designing Climate Change Mitigation Plans That Add Up," *Environmental Science & Technology*, vol. 47, no. 14, 2013: 8062–8069.

BAJŽELJ et al. (2014)
BAJŽELJ, Bojana et al. "Importance of food-demand management for climate mitigation," *Nature Climate Change*, vol. 4, no. 10, 2014: 924–929.

BARNOSKY et al. (2012)
BARNOSKY, Antony D. et al. "Approaching a state shift in Earth's biosphere," *Nature*, vol. 486, June 7, 2012: 52–58.

BASTIN et al. (2019)
BASTIN, Jean-Francois et al. "The global tree restoration potential," *Science*, vol. 365, July 5, 2019: 76–79.

BP (2018)
BP, 2018. *BP Statistical Review of World Energy 2018*, retrieved July 25, 2020, from: https://www.bp.com/content/dam/bp/business-sites/en/global/corporate/pdfs/energy-economics/statistical-review/bp-stats-review-2018-full-report.pdf.

BP (2019)
BP, 2019. *BP Statistical Review of World Energy 2019*, retrieved May 14, 2020, from: https://www.bp.com/en/global/corporate/energy-economics/statistical-review-of-world-energy.html.

CAPRA / LUISI (2016)
CAPRA, Fritjof, LUISI, Pier Luigi. *The Systems View of Life. A Unifying Vision*, Cambridge: Cambridge University Press, 2016[6].

CDP CARBON MAJORS REPORT (2017)
THE CARBON MAJORS DATABASE, 2017. GRIFFIN, Paul. *CDP Carbon Majors Report 2017*, retrieved May 14, 2020, from: https://climateaccountability.org/pdf/CarbonMajorsRpt2017%20Jul17.pdf.

CEBALLOS / EHRLICH / RAVEN (2020)
CEBALLOS, Gerardo, EHRLICH, Paul R., RAVEN, Peter H. "Vertebrates on the brink as indicators of biological annihilation and the sixth mass extinction," *Proceedings of the National Academy of Sciences of the United States of America* (PNAS), vol. 117, no. 24, 2020: 13596–13602.

COOK et al. (2013)
COOK, John et al. "Quantifying the consensus on anthropogenic global warming in the scientific literature," *Environmental Research Letters*, vol. 8, no. 2, 2013: 024024 (1–7).

COUMOU / RAHMSTORF (2012)
COUMOU, Dim, RAHMSTORF, Stefan. "A decade of weather extremes," *Nature Climate Change*, vol. 2, no.7, 2012: 491–496.

CROWTHER et al. (2015)
CROWTHER, Thomas W. et al. "Mapping tree density at a global scale," *Nature*, vol. 525, Sept. 2, 2015: 201–205.

EAT-LANCET SUMMARY (2019)
Summary Report of the EAT-*Lancet* Commission. *Food Planet Health. Healthy Diets from Sustainable Food Systems,* retrieved Apr. 19, 2020, from: https://eatforum.org/eat-lancet-commission/eat-lancet-commission-summary-report/. For the full commission see: **WILLETT / ROCKSTRÖM et al. (2019).**

EESI (2019)
ENVIRONMENTAL AND ENERGY STUDY INSTITUTE, 2019. *Fossil Fuel Subsidies: A Closer Look at Tax Breaks and Societal Costs (Fact Sheet)*, retrieved Feb. 8, 2020, from: https://www.eesi.org/files/FactSheet_Fossil_Fuel_Subsidies_0719.pdf.

ELLEN MacARTHUR FOUNDATION (2017)
ELLEN MacARTHUR FOUNDATION, 2017. *A new textiles economy: Redesigning fashion's future*, retrieved May 12, 2020, from: http://www.ellenmacarthurfoundation.org/publications.

ELLEN MacARTHUR FOUNDATION (2019)
ELLEN MacARTHUR FOUNDATION, 2019. *Urban Mobility System*, retrieved May 21, 2020, from: https://www.ellenmacarthurfoundation.org/assets/downloads/Mobility_All_Mar19.pdf.

EUROPEAN COMMISSION (2019)
EUROPEAN COMMISSION, 2019. *The European Green Deal. COM/2019/640 final,* retrieved Apr. 17, 2020, from: https://eur-lex.europa.eu/resource.html?uri=cellar:b828d165-1c22-11ea-8c1f-01aa75ed71a1.0002.02/DOC_1&format=PDF.

EUROPEAN COMMISSION (2020)
EUROPEAN COMMISSION, 2020. *European Climate Law. COM/2020/80 final*, retrieved Apr. 17, 2020, from: https://ec.europa.eu/info/sites/info/files/commission-proposal-regulation-european-climate-law-march-2020_en.pdf.

FAO (2011)
FOOD AND AGRICULTURE ORGANIZATION OF THE UNITED NATIONS, 2011. *Global food losses and food waste. Extent, causes and prevention*, Rome, retrieved Apr. 26, 2020, from: http://www.fao.org/3/a-i2697e.pdf.

FAO / IFAD (2019)
FOOD AND AGRICULTURE ORGANIZATION OF THE UNITED NATIONS / INTERNATIONAL FUND FOR AGRICULTURAL DEVELOPMENT, 2019. *United Nations Decade of Family Farming 2019–2028. Global Action Plan*, Rome, retrieved Apr. 19, 2020, from: http://www.fao.org/3/ca4672en/ca4672en.pdf.

FAO / IFAD / UNICEF / WFP / WHO (2020)
FOOD AND AGRICULTURE ORGANIZATION OF THE UNITED NATIONS / INTERNATIONAL FUND FOR AGRICULTURAL DEVELOPMENT / UNITED NATIONS CHILDREN'S FUND / WORLD FOOD PROGRAMME / WORLD HEALTH ORGANIZATION, 2020. *The State of Food Security and Nutrition in the World 2020. Transforming food systems for affordable healthy diets*, Rome, retrieved July 31, 2020, from: http://www.fao.org/documents/card/en/c/ca9692en.

FOTHERGILL / SCHOLEY (2019)
FOTHERGILL, Alastair, SCHOLEY, Keith. *Our Planet*, London: Transworld Publishers, 2019.

FRIEDLINGSTEIN et al. (2019)
FRIEDLINGSTEIN, Pierre et al. "Global Carbon Budget 2019," *Earth System Science Data* (ESSD), vol. 11, no. 4, 2019: 1783–1838.

GREENPEACE (2017)
GREENPEACE, 2017. COOK, Gary, JARDIM, Elizabeth. *Guide to Greener Electronics*, retrieved May 12, 2020, from: www.greenpeace.org/usa/reports/greener-electronics-2017.

GRISCOM et al. (2017)
GRISCOM, Bronson W. et al. "Natural climate solutions," *Proceedings of the National Academy of Sciences of the United States of America* (PNAS), vol. 114, no. 44, 2017: 11645–11650.

HAZEN (2012)
HAZEN, Robert M. *The Story of Earth: The First 4.5 Billion Years, from Stardust to Living Planet,* New York: Penguin Books, 2012.

HAZEN (2019)
HAZEN, Robert M. *Symphony in C. Carbon and the Evolution of (Almost) Everything*, London: William Collins, 2019.

HOEGH-GULDBERG et al. (2019)
HOEGH-GULDBERG, Ove et al. "The human imperative of stabilizing global climate change at 1.5°C," *Science*, vol. 365, Sept. 20, 2019: 1263 (1–11).

IEA (2019)
INTERNATIONAL ENERGY AGENCY, 2019. *IEA Key World Energy Statistics 2019*, retrieved May 14, 2020, from: https://www.iea.org/reports/key-world-energy-statistics-2019.

IPCC (2014)
INTERGOVERNMENTAL PANEL ON CLIMATE CHANGE, 2014. *Climate Change 2014: Synthesis Report. Contribution of Working Groups I, II and III to the Fifth Assessment Report of the Intergovernmental Panel on Climate Change*, Geneva, retrieved May 14, 2020, from: https://www.ipcc.ch/report/ar5/syr/.

IPCC (2018)
INTERGOVERNMENTAL PANEL ON CLIMATE CHANGE, 2018. Summary for Policymakers. In: *Global warming of 1.5°C*. World Meteorological Organization, Geneva, retrieved May 14, 2020, from: https://www.ipcc.ch/sr15/.

IPCC (2019/1)
INTERGOVERNMENTAL PANEL ON CLIMATE CHANGE, 2019. Summary for Policymakers. In: *Climate Change and Land: An IPCC Special Report on climate change, desertification, land degradation, sustainable land management, food security, and greenhouse gas fluxes in terrestrial ecosystems*, retrieved May 14, 2020, from: https://www.ipcc.ch/srccl/chapter/summary-for-policymakers/.

IPCC (2019/2)
INTERGOVERNMENTAL PANEL ON CLIMATE CHANGE, 2019. Summary for Policymakers. In: *IPCC Special Report on the Ocean and Cryosphere in a Changing Climate*, retrieved July 27, 2020, from: https://www.ipcc.ch/site/assets/uploads/sites/3/2019/11/03_SROCC_SPM_FINAL.pdf.

IPBES (2019)
INTERGOVERNMENTAL SCIENCE-POLICY PLATFORM ON BIODIVERSITY AND ECOSYSTEM SERVICES, 2019. *Summary for policymakers of the IPBES global assessment report on biodiversity and ecosystem services*, Bonn, retrieved May 14, 2020, from: https://www.ipbes.net/system/tdf/spm_global_unedited_advance.pdf?file=1&type=node&id=35245.

IMF (2019)
INTERNATIONAL MONETARY FUND, 2019. COADY, David, PARRY, Ian, LE, Nghia-Piotr, SHANG, Baoping. *Global Fossil Fuel Subsidies Remain Large: An Update Based on Country-Level Estimate*, retrieved Feb. 8, 2020, from: www.imf.org/~/media/Files/Publications/WP/2019/WPIEA2019089.ashx.

KEELING (1960)
KEELING, Charles D. "The concentration and isotopic abundances of carbon dioxide in the atmosphere," *Tellus*, vol. 12, no. 2, 1960: 200–203.

KLEIN (2015)
KLEIN, Naomi. *This Changes Everything. Capitalism vs. the Climate*, London: Penguin Books, 2015.

KLEIN (2019)
KLEIN, Naomi. *On Fire. The Burning Case for a Green New Deal*, London: Allen Lane, 2019.

KROMP-KOLB/FORMAYER (2018)
KROMP-KOLB, Helga, FORMAYER, Herbert. *+2 Grad. Warum wir uns für die Rettung der Welt erwärmen sollten*. Wien – Graz: Molden Verlag, 2018.

LANDRIGAN et al. (2018)
LANDRIGAN, Philip J. et al. "The Lancet commission on pollution and health," *The Lancet*, vol. 391, Feb. 3, 2018: 462–512.

LENTON/ROCKSTRÖM/GAFFNEY/RAHMSTORF/RICHARDSON/STEFFEN/SCHELLNHUBER (2019)
LENTON, Timothy M., ROCKSTRÖM, Johan, GAFFNEY, Owen, RAHMSTORF, Stefan, RICHARDSON, Katherine, STEFFEN, Will, SCHELLNHUBER, Hans Joachim. "Climate tipping points—too risky to bet against (Comment)," *Nature*, vol. 575, Nov. 28, 2019: 592–595, retrieved July 26, 2020, from: https://media.nature.com/original/magazine-assets/d41586-019-03595-0/d41586-019-03595-0.pdf.

LE QUÉRÉ et al. (2018)
LE QUÉRÉ, Corinne et al. "Global Carbon Budget 2018," *Earth System Science Data* (ESSD), vol. 10, no. 4, 2018: 2141–2194.

LOVELOCK/MARGULIS (1974)
LOVELOCK, James, MARGULIS, Lynn. "Atmospheric homeostasis by and for the biosphere: the Gaia hypothesis," *Tellus*, vol. 26, no. 1–2, 1974: 2–10.

LOVINS (2011)
LOVINS, Amory B., ROCKY MOUNTAIN INSTITUTE. *Reinventing Fire: Bold Business Solutions for the New Energy Era*, White River Junction: Chelsea Green Publishing, 2011.

MARCOTT et al. (2013)
MARCOTT, Shaun A. et al. "A Reconstruction of Regional and Global Temperature for the Past 11,300 Years," *Science*, vol. 339, Mar. 8, 2013: 1198–1201.

McDONOUGH (2016)
McDONOUGH, William. "Carbon is not the enemy," *Nature*, vol. 539, Nov. 17, 2016: 349–351.

McDONOUGH/BRAUNGART (2002)
McDONOUGH, William, BRAUNGART, Michael. *Cradle to Cradle: Remaking the Way We Make Things*, New York: North Point Press, 2002.

MEADOWS/MEADOWS/RANDERS/BEHRENS (1972)
MEADOWS, Dennis L., MEADOWS, Donella H., RANDERS, Jorgen, BEHRENS, William W. *The Limits to Growth: A Report for the Club of Rome's Project on the Predicament of Mankind*, New York: Universe Books, 1972. A PDF of the book is downloadable from: https://collections.dartmouth.edu/content/deliver/inline/meadows/pdf/meadows_ltg-001.pdf.

MULVANEY (2004)
MULVANEY, Robert. "How are past temperatures determined from an ice core?," *Scientific American* online, Sept. 20, 2004, retrieved May 18, 2020, from: https://www.scientificamerican.com/article/how-are-past-temperatures/.

OECD/IEA (2019)
ORGANIZATION FOR ECONOMIC CO-OPERATION AND DEVELOPMENT/INTERNATIONAL ENERGY AGENCY, 2019. *Update on recent progress in reform of inefficient fossil-fuel subsidies that encourage wasteful consumption*, retrieved May 18, 2020, from: https://oecd.org/fossil-fuels/publication/OECD-IEA-G20-Fossil-Fuel-Subsidies-Reform-Update-2019.pdf.

PETIT et al. (1999)
PETIT, Jean-Robert et al. "Climate and atmospheric history of the past 420,000 years from the Vostok ice core, Antarctica," *Nature*, vol. 399, June 3, 1999: 429–436.

POWELL (2015)
POWELL, James Lawrence. "Climate Scientists Virtually Unanimous: Anthropogenic Global Warming Is True," *Bulletin of Science, Technology & Society*, vol. 35 (5–6), 2015: 121–124.

RAHMSTORF/SCHELLNHUBER (2018)
RAHMSTORF, Stefan, SCHELLNHUBER, Hans Joachim. *Der Klimawandel*, Munich: C. H. Beck, 2018[8].

REN21 HIGHLIGHTS 2018 (2019)
RENEWABLE ENERGY POLICY NETWORK FOR THE 21st CENTURY, 2019. *Highlights of the REN21 Renewables 2018 Global Status Report in perspective*, retrieved May 18, 2020, from: https://

www.ren21.net/wp-content/uploads/2019/08/
Highlights-2018.pdf.

RICARDO/SZOSTAK (2009)
RICARDO, Alonso, SZOSTAK, Jack W. "Life on Earth. Fresh clues hint at how the first living organisms arose from inanimate matter," *Scientific American*, Sept. 2009: 54–61.

RIFKIN (2019)
RIFKIN, Jeremy. *The Green New Deal: Why the Fossil Fuel Civilization Will Collapse by 2028, and the Bold Economic Plan to Save Life on Earth*, New York: St. Martin's Press, 2019.

ROCKSTRÖM et al. (2009)
ROCKSTRÖM, Johan *et al.* "Planetary boundaries: Exploring the safe operating space for humanity," *Ecology and Society*, vol. 14, no. 2, 2009: 32–63.

ROCKSTRÖM/KLUM (2015)
ROCKSTRÖM, Johan, KLUM, Mattias. *Big World, Small Planet. Abundance Within Planetary Boundaries*, New York: Yale University Press, 2015.

ROCKSTRÖM et al. (2017)
ROCKSTRÖM, Johan *et al.* "A roadmap for rapid decarbonization," *Science*, vol. 355, Mar. 24, 2017: 1269–1271.

ROGELJ et al. (2019)
ROGELJ, Joeri *et al.* "Estimating and tracking the remaining carbon budget for stringent climate targets," *Nature*, vol. 571, July 17, 2019: 335–342.

ROMM (2018)
ROMM, Joseph J. *Climate Change. What Everyone Needs to Know*, New York: Oxford University Press, 2018[2].

SCHELLNHUBER et al. (2012)
SCHELLNHUBER, Hans Joachim *et al. Turn Down the Heat: Why a 4 °C Warmer World Must be Avoided*, Report for The World Bank, Nov. 2012, retrieved May 18, 2020, from: http://documents.worldbank.org/curated/en/865571468149107611/Turn-down-the-heat-why-a-4-C-warmer-world-must-be-avoided.

SCHELLNHUBER (2015)
SCHELLNHUBER, Hans Joachim. *Selbstverbrennung*, Munich: C. Bertelsmann Verlag, 2015.

SMIL (2011)
SMIL, Vaclav. "Harvesting the Biosphere: The Human Impact," *Population and Development Review*, vol. 37, no. 4, Dec. 2011: 613–636.

SMIL (2013)
SMIL, Vaclav. *Harvesting the Biosphere: What We Have Taken from Nature*, Cambridge: MIT Press, 2013.

STEFFEN/CRUTZEN/McNEILL (2007)
STEFFEN, Will, CRUTZEN, Paul J., McNEILL, John R. "The Anthropocene: Are Humans Now Overwhelming the Great Forces of Nature?," *Ambio,* vol. 36, no. 8, Dec. 2007: 614–621.

STEFFEN et al. (2015)
STEFFEN, Will *et al.* "Planetary boundaries: Guiding human development on a changing planet," *Science*, vol. 347, Feb. 13, 2015: 1259855 (1–10).

STEFFEN/BROADGATE/DEUTSCH/GAFFNEY/LUDWIG (2015)
STEFFEN, Will, BROADGATE, Wendy, DEUTSCH, Lisa, GAFFNEY, Owen, LUDWIG, Cornelia. "The trajectory of the Anthropocene: The Great Acceleration," *The Anthropocene Review*, vol. 2, no. 1, Apr 1, 2015: 81–98.

STEFFEN et al. (2018)
STEFFEN, Will *et al.* "Trajectories of the Earth System in the Anthropocene," *Proceedings of the National Academy of Sciences of the United States of America* (PNAS), vol. 115, no. 33, 2018: 8252–8259.

STIGLITZ (2019)
STIGLITZ, Joseph E. *People, Power, and Profits. Progressive Capitalism for an Age of Discontent*, London: Allen Lane, 2019.

STIPS et al. (2016)
STIPS, Adolf *et al.* "On the causal structure between CO_2 and global temperature," *Scientific Reports*, vol. 6, Article number: 21691, 2016.

STOKNES (2014)
STOKNES, Per Espen. "Rethinking climate communications and the 'psychological climate paradox,'" *Energy Research & Social Science*, vol. 1, March 2014: 161–170.

THUNBERG (2019)
THUNBERG, Greta. *No One is Too Small to Make a Difference*, London: Penguin Books, 2019.

TOLLEFSON (2014)
TOLLEFSON, Jeff. "The Case of the Missing Heat," *Nature*, vol. 505, Jan. 16, 2014: 276–278.

TOWNSEND/HOWARTH (2010)
TOWNSEND, Alan R., HOWARTH, Robert W. "Fixing the global nitrogen problem," *Scientific American*, Feb. 2010: 32–39.

UN (2019)
UNITED NATIONS, Department of Economic and Social Affairs, Population Division, 2019. *World Population Prospects: Highlights*, retrieved July 20, 2020, from: https://population.un.org/wpp/Publications/Files/WPP2019_Highlights.pdf.

UNEP (2019)
UNITED NATIONS ENVIRONMENT PROGRAMME, 2019. *Emissions Gap Report 2019. Executive Summary*, retrieved May 18, 2020, from: https://unepdtu.org/publications/emissions-gap-report-2019-executive-summary/.

UNFCCC (2015)
UNITED NATIONS FRAMEWORK CONVENTION ON CLIMATE CHANGE, 2015. *Paris Agreement*, retrieved May 18, 2020, from: https://unfccc.int/files/essential_background/convention/application/pdf/english_paris_agreement.pdf.

VAN KLINK et al. (2020)
VAN KLINK, Roel *et al.* "Meta-analysis reveals declines in terrestrial but increases in freshwater insect abundances," *Science*, vol. 368, Apr. 24, 2020: 417–420.

VON WEIZSÄCKER/WIJKMAN (2018)
VON WEIZSÄCKER, Ernst Ulrich, WIJKMAN, Anders. *Come On! Capitalism, Short-termism, Population and the Destruction of the Planet*, New York: Springer, 2018.

WACKERNAGEL/BEYERS (2019)
WACKERNAGEL, Mathis, BEYERS, Bert, GLOBAL FOOTPRINT NETWORK. *Ecological Footprint: Managing Our Biocapacity Budget*, Gabriola Island: New Society Publishers, 2019.

WARD/BROWNLEE (2000)
WARD, Peter D., BROWNLEE, Donald E. *Rare Earth: Why Complex Life Is Uncommon in the Universe*, New York: Corpernicus Books, 2000.

WEBSTER (2017)
WEBSTER, Ken. *The Circular Economy: A Wealth of Flows*, Cowes, Isle of Wight: Ellen MacArthur Foundation Publishing, 2017[2].

WEF (2019/1)
WORLD ECONOMIC FORUM, 2019. *Meat: The Future Series. Alternative Proteins*, retrieved May 18, 2020, from: https://www.weforum.org/whitepapers/meat-the-future-series-alternative-proteins.

WEF (2019/2)
WORLD ECONOMIC FORUM, 2019. *The Speed of the Energy Transition. Gradual or Rapid Change?*, retrieved May 18, 2020, from: https://www.wefo-

rum.org/whitepapers/the-speed-of-the-energy-transition.

WILLETT / ROCKSTRÖM *et al.* (2019)
WILLETT, Walter, ROCKSTRÖM, Johan *et al.*
"Food in the Anthropocene: the EAT-*Lancet*
Commission on healthy diets from sustainable
food systems," *The Lancet*, vol. 393, Feb. 2, 2019:
447–492.

WWF (2018)
WORLD WILDLIFE FUND, 2018. GROOTEN,
Monique, ALMOND, Rosamunde. *Living Planet
Report. 2018: Aiming Higher*, Gland, Switzerland,
retrieved July 26, 2020, from: https://c402277.ssl.
cf1.rackcdn.com/publications/1187/files/original/
LPR2018_Full_Report_Spreads.pdf.

Online Resources /
Open Access Data

**Climate Change / Carbon Emissions:
Up-to-date Status and Reports**

▪ Intergovernmental Panel on Climate Change
(IPCC): https://ipcc.ch > Reports

▪ World Meteorological Organization (WMO):
https://library.wmo.int/index.php?lvl=categ_
see&id=10959&main=1#.XsL2PS1XbOQ

▪ National Oceanic and Atmospheric Administration (NOAA): https://www.climate.gov/maps-data-
#global-climate-dashboard

▪ National Aeronautics and Space Administration
(NASA): https://climate.nasa.gov

▪ Global Carbon Project/CDIAC:
https://www.globalcarbonproject.org

▪ Climate Action Tracker:
https://climateactiontracker.org/countries/

Ecological Footprint

▪ Global Footprint Network:
http://data.footprintnetwork.org/
(open access country footprint data)
https://www.footprintcalculator.org/
(Footprint calculator)

World Energy

▪ International Energy Agency:
https://www.iea.org/data-and-statistics

▪ BP: https://www.bp.com/en/global/corporate/
energy-economics/statistical-review-of-world-
energy.html

World and World Development Statistics

▪ Our World in Data: https://ourworldindata.org

▪ World Bank: https://data.worldbank.org

▪ Gapminder: https://www.gapminder.org/
tools/#$chart-type=bubbles

For links to regularly updated statistics and new
scientific literature visit **www.now-jetzt.org**.

Index

Acknowledgments

The contents and aesthetics of this book have been shaped to a good part by the many conversations I had about them with friends and colleagues. First and foremost I thank Brigitte Felderer, whose critical reading of and enlightening remarks on the first draft led to a substantial expansion and rewrite of this book. I sincerely thank Mathis Wackernagel of Global Footprint Network, who very forthcomingly pointed me in the right direction regarding use of the Ecological Footprint concept. Apollina Smaragd and Michael Strasser accompanied the whole writing process with many helpful comments—so did Gabrielle Cram and Silvia Jaklitsch of Verlag für moderne Kunst, Sonja Gruber, Barbara Horvath, Kerstin Hosa, Simona Koch, Nadina Müllner, Veronika Rudorfer, Sonja Russ, and Austin Settle. Wolfgang Astelbauer engaged with the book's contents with far more dedication than his meticulous proofreading required him to do. Special thanks are due to Nina Sponar, who not only explored with me the many issues touched upon in these pages but also allowed me to use her as a voice in this book. She assisted me in innumerable ways during the long process of writing and designing this publication.

I sincerely thank Greenpeace for letting me use more than a dozen of their fantastic photos in this publication. For his forthcoming help with image rights and permissions I especially thank Mitja Kobal of Greenpeace Austria/CEE. Special thanks are also due to Bert Ulrich and his team at NASA, another great source of imagery in the book. Claudia Kettner-Marx of WIFO helped me with studies conducted by her institute on the mobility sector in Austria.

As far as materials for and the printing of this book are concerned I sincerely thank Bettina Wobetzky of Arctic Paper, who supported its production not only with great enthusiasm but also materially. Reinhard Gugler of Gugler GmbH I thank for his professional and friendly guidance regarding Cradle-to-cradle-certified printing. I am grateful for many years of enjoyable collaboration, creative inputs, and positive feedback to Hannes Fauland, Manfred Kostal, Sonja Russ, and especially Sandra Schmidt.

How this book was printed

This book was printed by Gugler GmbH in Melk, Austria, and meets the highest standards for eco-friendly printing. Gugler is *Cradle-to-cradle* (C2C)-certified, which attests to rigorous application of circular economy principles in a business' operations. The C2C-certificate (Silver) that Gugler was issued prescribes that:
* All paper must come from certified sustainable forestry (FSC and PEFC certificates).
* All other components used in the production process, such as printing ink, must surpass all the requirements of the Austrian Ecolabel.
* *Climate-positive* production is required, meaning that CO_2 emissions generated during production are compensated for at a rate of 110%. The generated proceeds directly benefit a reforestation project of the University of Natural Resources and Life Sciences, Vienna.
* Chemicals typically used in the production process, which have to be kept within restricted limits, are not used *at all*. No substances that might be harmful during bio-degradation are used.
* Energy efficiency of the production process is not sufficient: the use of renewable energies is required.
* Rigorous water stewardship has to be realized.
* Fair working conditions have to be realized.

The paper used for this book (Munken Polar by Arctic Paper) is Cradle-to-cradle-certified as well (Bronze). As of August 2020, so are all Munken Book and Design Papers, as well as Amber Graphic and Munken Kraft produced at the Arctic Paper Munkedals mill in Sweden. Arctic Paper Munkedals has been one of the most eco-friendly paper mills worldwide for years, but now it has become the first to successfully complete C2C-certification for its whole portfolio of products.

Planting 3 trees with this book

Beyond the eco-friendly C2C-certified production of this book, for every copy sold we will plant 3 trees via the Trillion Tree Campaign initiated by Plant for the Planet (www.plant-for-the-planet.org). You can track how many trees we have planted by going to www.trilliontreecampaign.org/explore and then zooming in on Europe > Austria > Vienna, and finding *NOW-JETZT*.
We hope to plant the 3,000 trees, which correspond to the first print run of 1,000 copies, as soon as possible. Hopefully, we will be able to plant more trees with future editions of the book :)

CERTIFIED
cradle to cradle
SILVER

Except binding
www.gugler.at

greenprint*
carbon positive printed

FSC
www.fsc.org
MIX
Paper from
responsible sources
FSC® C005108

✪ MUNKEN

printed according to the Guideline „Printed products" of the Austrian Eco-label, Gugler GmbH, Eco-Label No. 609, www.gugler.at

Publisher's note

Concept/texts/editor: Christian Schienerl
Proofreading: Wolfgang Astelbauer
Graphic design: Christian Schienerl,
Nina Sponar, intern: Meskerem Astelbauer,
SCHIENERL D/AD, Vienna
Lithography: Pixelstorm, Vienna
Printed by: Gugler GmbH, Melk
Paper: Munken Polar, 150 and 300 gsm
Print run: 600 German and 400 English copies
Fonts: Helvetica Now, Tatiana

Published by:
VfmK Verlag für moderne Kunst GmbH
Schwedenplatz 2/24
A-1010 Vienna
hello@vfmk.org
www.vfmk.org

ISBN 978-3-903796-02-7

Distribution
Europe: LKG, www.lkg-va.de
UK: Cornerhouse Publications,
www.cornerhousepublications.org
USA: D.A.P., www.artbook.com

Bibliographic information published by the
Deutsche Nationalbibliothek
The Deutsche Nationalbibliothek lists this publication in the Deutsche Nationalbibliografie; detailed
bibliographic data are available in the Internet at
dnb.de.

www.now-jetzt.org

Image credits
All images (photos and graphics) in this publication
are credited in captions next to the images.
All images are copyright-protected by the stated
photographers/authors, with all rights reserved.
For all images licensed as creative commons (CC)
the respective CC license is stated in brackets.
Some images used in this publication are in the
public domain and/or free to use for educational
and editorial purposes.
Despite the editor's best efforts, some photos
could not be traced to their sources and are
marked as "photo: source unknown." If you can
contribute information on those images' sources,
please contact the editor:
now-jetzt@schienerl.com

Front and back cover photos: NASA